# Essential Concepts of Remote Sensing

# Essential Concepts of Remote Sensing

Edited by **Matt Weilberg**

New York

Published by Callisto Reference,
106 Park Avenue, Suite 200,
New York, NY 10016, USA
www.callistoreference.com

**Essential Concepts of Remote Sensing**
Edited by Matt Weilberg

International Standard Book Number: 978-1-63239-317-3 (Hardback)

Printed in the United States of America.

# Contents

# Preface

The world is advancing at a fast pace like never before. Therefore, the need is to keep up with the latest developments. This book was an idea that came to fruition when the specialists in the area realized the need to coordinate together and document essential themes in the subject. That's when I was requested to be the editor. Editing this book has been an honour as it brings together diverse authors researching on different streams of the field. The book collates essential materials contributed by veterans in the area which can be utilized by students and researchers alike.

This book includes essential concepts in the application of remote sensing. Remote sensing has majorly profited almost all areas of human activity and development by providing a common platform to physical, natural and social activities for interaction and advancement. This book discusses the impacts of remote sensing on various areas of science, human activity and technology by presenting a selected number of high quality contributions related to various remote sensing applications organized under sections namely, Remote Sensing in Oceans & Cryosphere and Human Activity Assessment. The book includes contributions of prominent experts and researchers, who possess vast knowledge and years of experience in this field.

Each chapter is a sole-standing publication that reflects each author's interpretation. Thus, the book displays a multi-facetted picture of our current understanding of application, resources and aspects of the field. I would like to thank the contributors of this book and my family for their endless support.

Editor

# Section 1

## Oceans and Cryosphere

# Remote Sensing of Submerged Aquatic Vegetation

Hyun Jung Cho[1], Deepak Mishra[2,3] and John Wood[4]
*[1]Department of Integrated Environmental Science,*
*Bethune-Cookman University,*
*Daytona Beach, FL*
*[2]Department of Geosciences, Mississippi State University*
*[3]Northern Gulf Institute and Geosystems Research Institute,*
*Mississippi State University, MS State, MS*
*[4]Harte Research Institute for Gulf of Mexico Studies,*
*Texas A&M University-Corpus Christi,*
*Corpus Christi, TX*
*USA*

## 1. Introduction

Remote sensing has significantly advanced spatial analyses of terrestrial vegetation for various fields of science. The plant pigments, chlorophyll *a* and *b*, strongly absorb the energy in the blue (centered at 450 nm) and the red (centered at 670 nm) regions of the electromagnetic spectrum to utilize the light energy for photosynthesis. In addition, the internal spongy mesophyll structures of the healthy leaves highly reflect the energy in the near-infrared (NIR) (700- 1300) regions (Jensen, 2000; Lillesand et al., 2008). The distinctive spectral characteristics of the green plants, low reflectance in the visible light and high reflectance in NIR have have been used for mapping, monitoring and resource management of plants; and also have been used to develop spectral indices such as Simple Vegetation Index (SVI = NIR reflectance – red reflectance) and Normalized Difference Vegetation Index (NDVI = (NIR reflectance – red reflectance)/(NIR reflectance + red reflectance)) (Giri et al., 2007).

The simplicity and flexibility of vegetation indices allow comparison of data obtained under varying light conditions (Walters et al., 2008). NDVI was first suggested by Ruose et al. (1973) and is one of the earliest and most popular vegetation index used to date. It is usually applied in an attempt to decrease the atmospheric and surface Bidirectional Reflectance Distribution Function (BRDF) effects by normalizing the difference between the red and NIR reflectance by total radiation. Index values have been associated with various plant characteristics, including vegetation type (Geerken et al., 2005), vegetation cover (du Plessis, 1999), vegetation water content (Jackson et al., 2004), biomass and productivity (Fang et al., 2001), chlorophyll level (Wu et al., 2008), PAR absorbed by crop canopy (Goward & Huemmrich, 1992), and flooded biomass (Beget et al., 2007) at a broad span of scales from individual leaf areas to global vegetation dynamics.

## 2. Remote sensing of submerged aquatic vegetation (SAV)

### 2.1 Submerged aquatic vegetation (SAV)

Submerged aquatic vegetation (SAV) is a group of vascular plants that grow underwater which can grow to the surface of, but not emerge from shallow waters. SAV includes seagrass species that are a vital component of ecological processes, dynamics, and productivity of coastal ecosystems. Healthy beds of SAV provide nursery and foraging habitats for juvenile and adult fish and shellfish, protect them from predators, provide food for waterfowl and mammals, absorb wave energy and nutrients, produce oxygen and improve water clarity, and help settle suspended sediment in water by stabilizing bottom sediments (Jin, 2001; Findlay et al., 2006). Assessment of SAV distribution, composition, and abundance has been of a particular interest to coastal environmental managers, scientists, developers, and recreational users as the information serves as an excellent indicator of aquatic environmental quality.

### 2.2 Remote sensing of underwater habitats

Remote sensing is a valuable tool for monitoring benthic habitats such as SAV, benthic algae, and coral-reef ecosystems, and several researchers have tested airborne and space-borne sensor systems for such studies (e.g., Mishra et al., 2005). Spatial resolutions of these systems range from 30 m for the Landsat Thematic Mapper (TM) to 2.44 m for QuickBird multispectral data and 1 m or less for airborne hyperspectral data. Those evaluating the utility of TM have mapped subtidal coastal habitats (Khan et al., 1992), delineated sand bottoms (Michalek et al., 1993), classified coral reef zones (Mishra et al., 2005, 2006), evaluated the benthos (Matsunaga & Kayanne, 1997), and performed time series analyses (Dobson & Dustan, 2000). Similarly, researchers have used IKONOS (4 m) and QuickBird (2.44 m) imagery with radiative transfer models to map benthic habitats (Mishra et al., 2005, 2006) and apply a similar model to Airborne Imaging Spectroradiometer for Applications (AISA) hyperspectral data to identify benthic habitats (Mishra et al., 2007).

Mapping of SAV using satellite data has focused on supervised and unsupervised classifications based on signal variations in the multispectral bands, especially those in the short visible wavelengths with high water penetration (Ackleson & Klemas, 1987; Lyzenga, 1981; Marshall & Lee, 1994; Maeder et al., 2002; Ferguson & Korfmacher, 1997; Pasqualini et al., 2005). The NIR region is seldom used due to its high attenuation in water. When SAV beds are dense, the water is clear, and depth and sediment relatively constant, fine-scale spectral variation is often overlooked during classification. In other cases, the radiative transfer model is used to correct the solar angle, atmospheric perturbation, substrate type, and depth, but requires extensive *in situ* measurements (Zimmerman & Mobley, 1997).

Most of the currently available radiative transfer models or physics-based models have been applied to map benthic features in relatively clear aquatic environments (i.e. relatively deep, pristine coral reefs or seagrass meadows) and do not adequately correct for the strong NIR absorption by water (Mumby et al., 1998; Holden & LeDrew, 2001; Holden & LeDrew, 2002; Ciraolo, 2006; Brando et al., 2009). However, NIR reflectance serves as the primary cue for discriminating vegetation type and as the critical component for the widely used vegetation indices.

## 2.3 Dilemmas in remote sensing of shallow aquatic system and SAV

Remote sensing of benthic habitats is complicated because of several factors including (1) atmospheric interferences, (2) variability in water depth, (3) water column attenuation, and (4) variability in bottom albedo or bottom reflectance. In the case of aquatic remote sensing, the total signal received at satellite altitude is dominated by radiance contributed by atmospheric scattering, and only 8-10% of the signal corresponds to the water reflectance and reflectance from benthic features (Kirk, 1994, Mishra et al., 2005). Therefore, it is advisable to correct for atmospheric effects to retrieve any quantitative information for surface waters or benthic habitats from satellite images. Therefore, the lack of a rigorous absolute atmospheric correction procedure can introduce significant errors to a satellite derived benthic habitat map. There is also a tendency among benthic mapping researchers to use a relative atmospheric correction procedure such as a deep-water pixel correction, especially when local aerosol data and validation data are lacking. This often causes mediocre classification results.

Knowledge of the optical properties of the water column can help eliminate changes in reflectance attributable to variable depth and water column attenuation effects, which often lead to misclassification of the benthos (Mishra et al., 2005). Mishra et al (2005) proved that to derive accurate bottom albedo or bottom reflectance using a radiative transfer model, water depth and water column optical properties (absorption and backscattering) should be known for the study area. Knowledge of optical bottom albedo for shallow waters is necessary to model the underwater and above-water light field, to enhance underwater object detection or imaging, and ultimately to determine the distribution of benthic habitats (Gordon & Brown, 1974). Mishra et al (2005, 2006, 2007) also point out that the signals measured by a sensor from above the water surface of a shallow marine environment are highly affected by phytoplankton abundance (chlorophyll absorption), water column interactions (absorption by water and scattering by suspended sediments), and radiance reflected from the bottom. For the bottom contribution to be retrieved by a sensor the water column contributions have to be removed and the optical properties have to be known or at least be derivable. However, it is very challenging to measure these optical properties accurately because of logistical issues and instrumentation errors, which also leads to an inaccurate benthic mapping project.

Variability in bottom types and hence albedo gives rise to a mixed spectral response that often reduces the classification accuracy. Specific problems such as complex benthic combinations (e.g., sandy areas with variable amounts of algal cover; variation in color, texture, size), and error in depth estimation can also have a considerable impact on the classification results. Mishra et al (2005) proposed several solutions to increase the number of elements separable by a classification scheme and the classification accuracies including an extensive field campaign acquiring substantial samples to enable statistical evaluation for each class and deriving detailed ecological and biological information for each *in situ* data point. Close-range hyperspectral studies that may aid in discriminating between different types of benthic features can be used to develop baseline spectra to help minimize spectral confusion in satellite imagery.

Shallow littoral areas (generally the areas between the shoreline to a water depth 2 m) are one of the most productive habitats, yet the most sensitive landscape to human-induced environmental alteration and global climate changes. Modeling of optical water properties for the littoral zone is more complicated due to rapidly changing water depth and/or substrate and higher amounts of Colored Dissolved Organic Matter (CDOM) and/or suspended particles (phytoplankton, seston, and inorganic particles) compared to the deeper portions of oceans. In addition, bottom backscattering in the shallow areas is more significant, which makes the NIR signals more important, especially in areas that contain substantial amount of seagrasses, benthic algae, or phytoplankton (Kutser et al., 2009) and that the conventional Beer-Lambert's exponential light attenuation with depth is not applicable (Holden & LeDrew, 2002).

Upwelling signals from water bodies contain several components including reflectance from water surface, water column (suspended matter), and bottom backscattering (SAV and substrate) (Spitzer & Dirks, 1987). The aforementioned conventional vegetation indices also are not effectively used for plants that grow underwater or that are temporarily flooded (Beget & Di Bella, 2007; Cho et al. 2008) because the water overlying the vegetation canopies reduces the vegetation effects of 'red absorption' and the 'NIR reflectance' (Han & Rundquist, 2003; Cho, 2007; Cho et al., 2008; Fig. 1). Differentiation of the SAV spectral signature from bare substrate or algae is further limited in shallow coastal waters that are more turbid than open ocean waters (Bukata, 1995) due to higher levels of phytoplankton, suspended sediment, and dissolved color. According to our on-going study using hyperspectral data obtained over both experimental tanks and field seagrass habitat, the SAV signal rapidly decreases as water depth increases, and almost completely disappears within a depth of 0.5 meter in even mildly turbid waters (turbidities > 12 NTU).

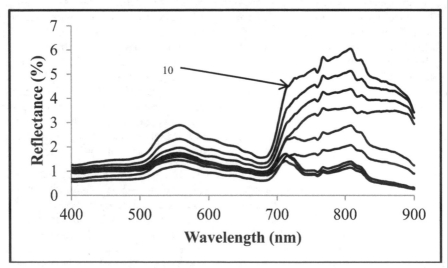

Fig. 1. Depth-induced reflectance variation of submerged aquatic vegetation (SAV) in clear water between 10 cm and 50 cm above the SAV canopy. The line for the highest reflectance is at 10 cm and the reflectance continuously decreases with water depth increases.

## 2.4 SAV mapping using hyperspectral data

Two decades ago, only spectral remote sensing experts had access to hyperspectral images or the software tools necessary to take advantage of such images. Over the past decade, hyperspectral image analysis has matured into one of the most powerful and fastest growing technologies in the field of remote sensing (Phinn et al., 2008). While multispectral remote sensing systems detect radiation in a small number of broad regions of the electromagnetic spectrum, hyperspectral sensors acquire numerous very narrow, contiguous spectral bands throughout the visible, near-infrared, mid-infrared, and thermal infrared portions of the electromagnetic spectrum for every pixel in the image, yielding much more detailed spectral data (Govender et al., 2009).

Collection and processing of hyperspectral imagery can be quite costly, depending on the size of the area to be studied. In order for the imagery to be usable for sub-aquatic analysis, the following guidelines are suggested by Finkbeiner et al. (2001):

- The best time of year for collecting hyperspectral imagery may occur in early summer, during the season of maximum biomass, and when there is less epiphytic coverage.
- The imagery should be collected when turbidity is low; this is often during times of low or no winds. High turbidity may also be caused by heavy rains, winds on previous days, and localized dredging. Often, boat traffic may cause a localized but far-spreading plume of turbidity, as sediments are re-suspended.
- Winds can also cause problems other than turbidity, such as wrack lines, debris lines, whitecaps, and areas with unacceptable amounts of glint. As a general rule, winds less than 8 kph are acceptable, winds between 8-15 kph may be acceptable depending on the locality, and winds higher than 15 kph are generally unacceptable.
- Tidal stage can play an important role in the success of imagery collection. Consult local and/or NOAA tide gauges to plan for acquisition within 2 hours of the lowest tide for the collection area, unless the estuary drains an area of highly turbid or tannic water, in which case, a rising tide may be desirable.
- Collection times should be planned to adjust for sun angle, to avoid both sun glint and shadows. As a general rule, sun angles between 30° and 45° are recommended; different sensors may allow or require more or less angle.
- Clouds and haze create areas of shadows and distortion as well as white or gray streaks in the imagery, and should be avoided as much as possible.
- Field work should occur simultaneously with the sensor flight. Since it is virtually impossible to collect all the field data needed for signature development and accuracy assessment in the same time frame as the flights occur, every effort should be taken to gather field measurements as close to the actual flight as possible, and under similar conditions.
- Field data should include measurements of reflectivity, turbidity, empirical or anecdotal data on epiphytic coverage, bottom type and reflectance, classification of the field point, and precise location. Locate these field measurements within a large enough patch that there will be no ambiguity, and consider the spatial sphere of uncertainty. For instance, if the imagery will have a radiometric accuracy of approximately 2 m, the location should be consistent out to a four meter radius.

The unique spectral signatures of vegetation are often used as training data for hyperspectral imagery classification. Chlorophyll and other pigments are found in SAV as in other photosynthesizing vegetation, however, the ratio of these to each other will differ by species, as well as with changes in conditions and stressors (Govender et al., 2009).

While these minor differences can be detected above the surface in spite of epiphytic coverage (Fyfe, 2003), detection of these differences below the surface may be hampered or dampened by the effects of the water column. Depth, water clarity or turbidity, organic and inorganic materials within the water column, the surface of the water itself, and physical properties such as the absorption of energy in the NIR and beyond can all affect the ability to discriminate the relatively small differences in ratios of accessory pigments and chlorophyll (Kutser 2004).

## 3. Case study in SAV mapping using hyperspectral data

### 3.1 Hyperspectral algorithm to correct overlying water effects

A new water-depth correction algorithm was developed to improve detection of underwater vegetation spectra signals. The algorithm was developed conceptually, calibrated, and validated using experimental and field data. The conceptual model was based on the idea that the upwelling signals measured from a water surface is the sum of the energy reflected from the water surface, the water column, and also from the water bottom surface. The energy reflected from the water surface and the water column (the volumetric reflectance) was combined as a single term because the surface reflectance is a constant and does not change with water depth changes (Lu & Cho, 2011).

The effects of the overlying water column on upwelling hyperspectral signals were modeled by empirically separating the energy absorbed and scattered by the water using data collected through a series of controlled experiments using hypothetical bottom surfaces that either 100% absorbs or 100% reflects (Cho & Lu, 2010). Later, the white surface (the 100% reflecting surface) was replaced with a gray surface with a known reflectance to reduce problems associated with enhanced multi-path scattering (Lu & Cho, 2011). The experimental setting allowed the calculation of water absorption and scattering values for up to a water depth of 60 cm, and the light remaining at water depths that were beyond the experimentally measured points were estimated by establishing the mathematical relationships between water depth and the vertical attenuation coefficient ($K_d$) derived from the experimental data (Washington et al., 2011). The depth- and wavelength-dependent water absorption and volumetric scattering factors (0-100 cm; 400 – 900 nm) were calculated and applied to independently-measured underwater vegetation signals and airborne hyperspectral data taken over shallow seagrass beds, to remove the effects of the overlying water.

### 3.2 Successful water correction in the infrared region

The empirically driven correction algorithm significantly restored the vegetation signals, especially in the NIR region, when applied to independently measured reflectance of underwater plants taken over indoor and outdoor tanks (Cho & Lu, 2010; Washington et al. 2011). The algorithm was also successful in restoring the NIR signals originating from seagrass-dominated sea floors when applied to airborne hyperspectral data of Mississippi and Texas coastal waters (Cho et al., 2009; Lu & Cho, 2011; Fig. 2). As stated earlier, NIR reflectance serves as the primary cue for discriminating benthic vegetation from other substrates. Due to the restored NIR reflectance, the correction algorithm increased the NDVI values for the seagrass pixels (Lu & Cho 2011).

### 3.3 Seagrass classification using water corrected image

### 3.3.1 Ground truth data collection

Several hundred ground data points were collected over seagrass beds in Redfish Bay, Texas, in the summer of 2008 (June – July). Seagrass species makeup, water depth, vegetation percent coverage, and bottom substrate type were recorded at each site. Site

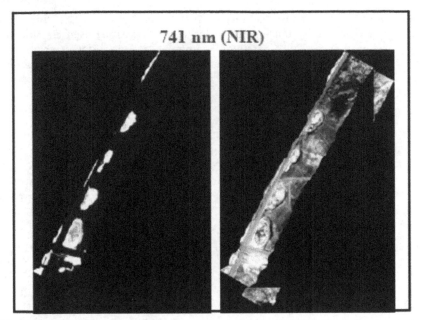

Fig. 2. The original (left) and water-corrected (right) airborne AISA Eagle hyperspectral image at 741 nm obtained over seagrass beds in Redfish Bay, TX in 2008.

location was recorded to accuracies within 1 m using a Real Time Kinetic (RTK) GPS. When necessary, the preselected random sites were shifted to avoid dry or unreachable locations.

The field collected data were entered into a spatial database along with descriptive attributes to help determine which class each sampling site would be assigned to. Data points were then randomly divided into training or accuracy assessment points.

### 3.3.2 Image processing and vector classification

Image data were obtained in 63 bands of the AISA Eagle Hyperspectral sensor over the seagrass beds in October 2008 and corrected for atmospheric effects. Since selection of the proper bands for analysis helps reduce noise introduction and processing burden (Borges et al. 2007), several selection techniques were used within this project, including Principal Component Analysis and regression analysis. Ultimately, seven bands recommended by Fyfe (2003) were used. To reduce noise, these were again reduced to 5 bands, as recommended by Cho et al. (2009).

Image segmentation is performed prior to image classification. Segmentation groups like pixels into homogenous areas. Initially, an unsupervised classification using ISODATA (Iterative Self Organized Data Analysis Technique A) (De Alwis et al., 2007) was performed. After the initial image classification, each segmented vector in the output was assigned to a seagrass/substrate class (i.e. *Thalassia testudinum*, *Halodule wrightii*, *Ruppia maritima*, Mixed Beds, Bare, or Unclassified). Those which contain only one type of point ('*Halodule*') were considered to be finally classified. Those classified as mixed or unknown were removed from the classified vector set, a mask created of their spatial footprint, and the entire classification procedure re-run on the image for that footprint area only. When it became impossible to further classify the image by this method, a supervised classification was performed, using a selection of training points as training data, which produced monospecific vectors. The same mixed-method classification procedure was performed to the image data after the water correction algorithm was applied.

### 3.3.3 Improved accuracy assessment

The water correction algorithm improved the classification accuracy results in an image subset from an overall accuracy of 28% to approximately 36%. Identification of the species *Halodule wrightii* improved from 33% user's accuracy to almost 78%, and *Thalassia testudimun* from 0% users accuarcy to almost 17%. Although these numbers appear to be somewhat low, several factors must be considered: this analysis only used a subset of the imagery, which allowed the area analysed to have less variation, but there was also less training and accuracy assessment data to work with.

Using the full set of imagery and the combination of supervised and unsupervised classification, results have improved considerably; we have achieved overall accuracies of over 60%. In addition, we have calculated seagrass presence/absence using the complete corrected imagery, achieving overall accuracies of over 95%. With the future addition of *in situ* turbidity and bathymetric data, these accuracies should continue to improve.

## 4. Current efforts in developing water correction module and graphical user interface

We have continued improving the algorithm by including turbidity (measured in NTU, Nephelometric Turbidity Unit) as an additional function. In addition, the water correction algorithm is currently being implemented as a module that can be called from the ENVI (ENvironment for Visualizing Images, ITT Visual Information Solutions)'s programming environment using a Graphical User Interface (GUI) (Gaye et al. 2011). Under the current module and the GUI, users are able to select water depth (0 – 2.0 m) and turbidity (0-20 NTU) using slide bars, for which (a) given hyperspectral band(s) can be corrected. The corrected reflectance can be generated and compared to the original one at either a given pixel, within a small subset; and the original and corrected images can be displayed.

## 5. Conclusion

Remote sensing has significantly advanced spatial analyses on terrestrial vegetation for various fields of science. However, mapping of benthic vegetation or submerged aquatic vegetation (SAV), using remotely sensed data is complicated due to several factors including atmospheric interferences, variability in water depth and bottom albedo, and water column attenuation by scattering and absorption. Hence, correction for the atmospheric and the overlying water column effects is necessary to retrieve any quantitative information for SAV from satellite and airborne images, especially when using hyperspectral data. Significant misclassification of the SAV often occurs due to the lack of information on *in situ* water depths and water column optical properties. Most of the currently available radiative transfer models only work well when applied to mapping of benthic features in relatively clear aquatic environments, but they do not correct for strong water absorption of the near infrared energy. The fluctuating water depths, high amounts of suspended particles and colored dissolved organic matter in shallow littoral zones make it even more challenging to map benthic vegetation using remotely sensed data. A new water-depth correction algorithm was developed conceptually, and calibrated and validated using experimental and field data. The effects of the overlying water column on upwelling hyperspectral signals were modeled by empirically separating the energy absorbed and scattered by the water using data collected through a series of controlled experiments. The empirically driven algorithm significantly restored the vegetation signals, especially in the NIR region. Due to the restored NIR reflectance, which serves as the primary cue for discriminating SAV from other substrates, use of the water corrected airborne data increased the NDVI values for the SAV pixels and also improved the seagrass classification accuracy. Our continuing efforts to incorporate turbidity and CDOM into the algorithm, in developing a graphical user interface, and in implementing the algorithm into a module that can be called from commercially available image processing software promise a user-friendly application and wide use of the algorithm in the near future.

## 6. Acknowledgment

The work was supported by grants from the National Geospatial-Intelligence Agency (Grant No. HM1582-07-1-2005 and HM1582-08-1-0049) and National Oceanic and Atmospheric Administration-Environmental Cooperative Sciences Center (Grant No.NA17AE1626).

## 7. References

Ackleson, S. G., & Klemas, V. (1987). Remote Sensing of Submerged Aquatic Vegetation in Lower Chesapeake Bay: A Comparison of Landsat MSS to TM imagery. *Remote Sensing of Environment*, Vol.22, No.2, pp. 235–248, ISSN 0034-4257

Beget, M. & Di Bella, C. (2007). Flooding: The Effects of Water Depth on Spectral Response of Grass Canopies. *Journal of Hydrology*, Vol.335, No.3-4, pp. 285-294. ISSN 0022-1694

Borges, J.; Marçal, A. & Dias, J. (2007). Evaluation of feature extraction and reduction methods for hyperspectral images. In: *New Developments and Challenges in Remote Sensing*, Bochenek, Z., pp. 265-284, Millpress, ISBN 978-90-5966-053-3, Rotterdam

Brando, V.E.; Anstee, J.M.; Wettle, M.; Dekker, A.G.; Phinn, S.R. & Roelfsema, C. (2009). A Physics Based Retrieval and Quality Assessment of Bathymetry from Suboptimal Hyperspectral Data. *Remote Sensing of Environment*, Vol.113, No.4, pp. 755-770, ISSN 0034-4257

Bukata, R.; Jerome, J.; Kondratyev, K. & Pozndyakov, D. (1995). Optical Properties and Remote Sensing of Inland Coastal Waters. New York: CRC Press.ISBN 978-0849347542 Cho, H.J. & Lu, D. (2010). A Water-depth Correction Algorithm for Submerged Vegetation Spectra. *Remote Sensing Letters*, Vol.1, No.1, pp. 29-35, ISSN 2150-7058

Cho, H.J., & Lu, D. (2010). A Water-depth Correction Algorithm for Submerged Vegetation Spectra. Remote Sensing Letters, Vol.1, No.1, pp. 29-35

Cho, H.J.; Lu, D. & Washington, M. (2009). Water Correction Algorithm Application for Underwater Vegetation Signal Improvement. *Proceedings of the 2009 MS Water Resource Conference*, pp. 154-157

Cho, H.J.; Kirui, P. & Natarajan, H. (2008). Test of Multi-spectral Vegetation Index for Floating and Canopy-forming Submerged Plants, *International Journal of Envirnmental research and Public Health*, Vol.5, No.5, pp. 447-483, ISSN 1661-7827

Cho, H. (2007). Depth-variant Spectral Characteristics of Submerged Aquatic Vegetation (SAV) detected by Landsat 7 ETM+. *International Journal of Remote Sensing*, Vo.28, No.7, pp. 1455-1467, ISSN 1366-5901

Ciraolo, G.; Cox, E.; La Loggia, G. & Maltese, A. (2006). The Classification of Submerged Vegetation Using Hyperspectral MIVIS Data. *Annals of Geophysics*, Vol. 49, No.1, pp. 287-294, ISSN 2037-416X

De Alwis, D.; Easton, Z., Dahlke, H., Philpot, W., & Steenhuis, T. (2007). Unsupervised Classification of Saturated Areas using a Time Series of Remotely Sensed Images. *Hydrology and Earth System Sciences* Vol 11, No. 511, pp. 1609-1620, ISSN 1027-5606

Dobson, E. & Dustan, P. (2000). The Use of Satellite Imagery for Detection of Shifts in Coral Reef Communities. *Proceedings, American Society for Photogrammetry and Remote Sensing*, Washington, DC, USA, May 22-26, 2002, (CD-ROM)

du Plessis, W.P. (1999). Linear Regression Relationships between NDVI, Vegetation and Rainfall in Etosha National Park, Namibia. *Journal of Arid Environments*, Vol.42, No.4, pp. 235-260, ISSN 0140-1963

Fang J.Y.; Piao S.L.; Tang Z.Y.; & Peng J.W. (2001). Interannual Variability in Net Primary Production and Precipitation. *Science*, Vol.29, No.5526, pp. 293:1723, ISSN 0036-8075

Ferguson, R.L.; & Korfmacher, K (1997). Remote Sensing and GIS Analysis of Seagrass Meadows in North Carolina, USA. *Aquatic Botany*, Vol.58, No. 3-4, pp. 241-258, ISSN 0304-3770

Findlay, S.E.G.; Nieder, W.C.; Blair, E.A. & Fischer, D.T. (2006). Multi-scale Controls on Water Quality Effects of Submerged Aquatic Vegetation in the Tidal Freshwater Hudson River. *Ecosystems*, Vol.9, No.1, pp. 84-96, ISSN 1432-9840

Finkbeiner, M.; Stevenson, B., & Seaman, R. (March 2001). *Guidance for Benthic Habitat Mapping: An Aerial Photographic Approach*. U.S. NOAA Coastal Services Center. http://www.csc.noaa.gov/benthic/mapping/pdf/bhmguide.pdf

Fyfe, S. (2003). Spatial and Temporal Variation in Spectral Reflectance: Are Seagrass Species Spectrally Distinct? *Limnology and Oceanography* Vol. 48, No. 1, pp. 464-479, ISSN 1939-5590

Gaye, G.; Kim, H.; Cho, H.J. (2011). Utilizing the Significant Spectral Bands for Image Outputs for SAV from the Water Effect Correction Module. Proceeding of the 2011 WORLDCOMP (CD-ROM)

Geerken, R.; Zaitchik, B.; & Evans, J.P. (2005). Classifying Rangeland Vegetation Type and Fractional Cover of Semi-arid and Arid Vegetation Covers from NDVI Time-series. *International Journal of Remote Sensing*, Vol.26 No.24, pp. 5535−5554, ISSN 1366-5901

Giri, C.; Pengra, B.; Zhu, Z.; Singh, A. & Tieszen, L. (2007). Monitoring Mangrove Forest Dynamics of the Sundarbans in Bangladesh and India Using Multitemporal Satellite Data from 1973-2000. *Estuarine, Coastal and Shelf Science*, Vol. 73, No.1-2, pp. 91-100. ISSN 0272-7714

Gordon, H.R. & Brown, O.B. (1974). Influence of Bottom Albedo on the Diffuse Reflectance of a Flat Homogeneous Ocean. *Applied Optics*, Vol.13, No.9, pp. 2153-2159, ISSN 1559-128X

Govender, M.; Chetty, K. & Bulcock, H. (2009). A Review of Hyperspectral Remote Sensing and Its Application in Vegetation and Water Resource Studies. *Water SA* Vol. 33, pp. 145-152, ISSN 1816-7950

Goward, S.N.; & Huemmrich, K.F. (1992). Vegetation Canopy PAR Absorptance and the Normalized Difference Vegetation Index: An Asssessment Using the SAIL model. *Remote Sensing of Environment*, Vol.39, No.2, pp. 119-140, ISSN 0034-4257

Han, L. & Rundquist, D.C. (2003). The Spectral Responses of Ceratophyllum demersum at Varying Depths in an Experimental Tank. International Journal of Remote Sensing, Vol. 24, pp. 859-864, ISSN 1366-5901

Holden, H. & LeDrew, E. (2001). Effects of the Water Column on Hyperspectral Reflectance of Submerged Coral Reef Features. Bulletin of Marine Science, Vol.69, No.2, pp. 685-699, ISSN 0007-4977

Holden, H. & LeDrew, E. (2002). Measuring and Modeling Water Column Effects on Hyperspectral Reflectance in a Coral Reef Environment. Remote Sensing of Environment, Vol. 81, No. 2-3, pp. 300-308, ISSN 0034-4257

Jackson, T.J.; Chen, D.; Cosh, M.; Li, F.; Anderson, M.; Walthall, C.; Doriaswamy, P. & Hunt, E.R. (2004). Vegetation Water Content Mapping Using Landsat Data Derived Normalized Difference Water Index for Corn and Soybeans. Remote Sensing Of Environment, Vol.92, No..3 pp. 475–482, ISSN 0034-4257

Jensen, J.R. (2000). Introductory Digital Image Processing: A Remote Sensing Perspective. Prentice Hall, Inc. Upper Saddle River, NJ 07458, ISBN-13: 978-0131453610, Upper Saddle River, NJ

Jin, X. (2001). Technologies of Lake Eutrophication Control management. Beijing: Chemistry Industry Press.

Khan, M.A.; Fadlallah, Y.H. & AL-Hinai, K.G. (1992). Thematic Mapping of Subtidal Coastal Habitats in the Western Arabian Gulf Using Landsat TM Data - Abu Ali Bay, Saudi Arabia. International Journal of Remote Sensing. Vol.13, No.4, pp. 605 – 614, ISSN 1366-5901

Kirk, J.T.O. (1994). Light and Photosynthesis in Aquatic Ecosystems. Cambridge University Press, ISBN 978-0-521-15175-7, Cambridge, UK

Kutser, T. (2004). Quantitative Detection of Chlorophyll in Cyanobacterial Blooms by Satellite Remote Sensing. Limnology and Oceanography Vol.49, No. 6, pp. 2179-2189, ISSN 1939-5590

Kutser, T.; Vahtmae, E. & Praks, J. (2009). A Sun Glint Correction Method for Hyperspectral Imagery Containing Areas with Non-negligible Water Leaving NIR Signal. Remote Sensing of Environment, Vol.113, No.110, pp. 2267-2274, ISSN 0034-4257

Lillesand, T.M., Kiefer, R.W., and Chipman, J.W. (2008). Remote Sensing and Image Interpretation. John Wiley and Sons, Inc., 111 River Street, Hoboken: NJ , ISBN-13 978-0470052457

Lu, D. & Cho, H.J. (2011). An Improved Water-depth Correction Algorithm for Seagrass Mapping Using Hyperspectral Data. Remote Sensing Letters, Vol.2, No.2, pp. 91-97, ISSN 2150-7058

Lyzenga, D. (1981). Remote Sensing of Bottom Reflectance and Water Attenuation Parameters in Shallow Water Using Aircraft and Landsat data. International Journal of Remote Sensing, Vol.2, No.1, pp. 71-82, ISSN 1366-5901

Maeder, J.; Narumalani, S.; Rundquist, D.C.; Perk, R.L.; Schalles, J.; Hutchins, K. & Keck. J. (2002). Classifying and Mapping General Coral-reef Structure Using Ikonos Data. Photogrammetric Engineering & Remote Sensing, Vol.68, No.12, pp. 1297-1305, ISSN 0099-1112

Marshall, T., & Lee, P. 1994. An Inexpensive and Lightweight Sampler for the Rapid Collection of Aquatic Macrophytes. Journal of Aquatic Plant Management, Vol.32, pp. 77-79, ISSN 0146-6623

Matsunaga, T. & Kayanne, H. (1997). Observation of Coral Reefs on Ishigaki Island, Japan, Using Landsat TM Images and Aerial Photographs, Proceedings of the Fourth

*International Conference on Remote Sensing for Marine and Coastal Environments*, Vol. I, pp. 657-666, ISSN 1066-3711, Orlando, Florida, USA, March 17-19, 1997

Michalek, J.; Wagner, T.; Luczkovich, J. & Stoffle, R. (1993). Multispectral Change Vector Analysis for Monitoring Coastal Marine Environments. *Photogrammetric Engineering and Remote Sensing*, Vol. 59, No.3, pp. 381-384, ISNN 0099-1112

Mishra, D., Narumalani, S.; Rundquist, D. & Lawson, M. (2005). High Resolution Ocean Color Remote Sensing of Benthic Habitats: A Case Study at the Roatan Island, Honduras. *IEEE Transactions in Geosciences and Remote Sensing*, Vol. 43, No.7, pp. 1592-1604, ISSN 0196-2892

Mishra, D., Narumalani, S.; Rundquist, D. & Lawson, M. (2006). Benthic Habitat Mapping in Tropical Marine Environments Using QuickBird Imagery. *Photogrammetric Engineering and Remote Sensing*, Vol.72, No.9, pp. 1037-1048, ISSN 0099-1112

Mishra, D., Narumalani, S.; Rundquist, D.; Lawson, M. & Perk, R. (2007). Enhancing the Detection and Classification of Coral Reef and Associated Benthic Habitats: A Hyperspectral Remote Sensing Approach. *Journal of Geophysical Research*, Vol.112, C08014, doi:10.1029/2006JC003892

Mumby, P.; Clark, C.; Green, E. & Edwards, A. (1998). Benefits of Water Column Correction and Contextual Editing for Mapping Coral Reefs. *International Journal of Remote Sensing*, Vol.19, No.1, pp. 203-210, ISSN 1366-6901

Pasqualini, V.; Pergent-Martini, C.; Pergent, G.; Agreil, M.; Skoufas, G.; Sourbes, L. & Tsirika, A. (2005). Use of SPOT 5 for Mapping Seagrasses: An Application to *Posidonia oceanica*. *Remote Sensing of Environment*, Vol.94, No.1, pp. 39-45, ISSN 0034-4257

Phinn, S.; Roelfsema, C.; Dekker, A.; Brando, V. & Anstee, J. (2008). Mapping Seagrass Species, Cover and Biomass in Shallow Waters: An Assessment of Satellite Multi-spectral and Airborne Hyper-spectral Imaging Systems in Moreton Bay (Australia). *Remote Sensing of Environment 2008*, Vol.112, No.8, pp. 3413-3425, ISSN ISSN 0034-4257

Rouse, J.; Haas, R.; Schell, J. & Deering, D. (1973). Monitoring Vegetation Systems in the Great Plains with ERTS. *3rd. ERTS Symposium*, NASA SP-351 I, 309-317

Spitzer, D. & Dirks, R. (1987). Solar Radiance Distribution in Deep Nataural Waters Including Fluorescence Effects. *Applied Optics*, Vol. 26, Issue 12, pp. 2427-2430, ISSN 1559-128X

Walters, B.; Ronnback, P.; Kovacs, J., Crona, B., Hussain, S., Badola, R., Primavera, J.H., Barbier, E. & Dahdouh-Guebas, F. (2008). Ethnobiology, Socio-economics and Adaptive Management of Mangroves: A Review. *Aquatic Botany*, Vol. 89, No.2, pp. 220–236, ISSN 0304-3770

Washington, M.; Kirui, P.; Cho, H. & Wafo-Soh, C. (2011). Data-driven Correction for Light Attenuation in Shallow Waters. *Remote Sensing Letters*, Vol.3, No.4, pp. 335-342, ISSN 2150-7058

Wu, C.; Niu, Z.; Tang, Q. & Huang, W. (2008). Estimating Chlorophyll Content from Hyperspectral Vegetation Indices: Modeling and Validation. *Agricultural and Forest Meteorology*, Vol. 48, No. 8-9, pp. 1230-1241, ISSN 0168-1923

Zimmerman, R. & Mobley, C. (1997). Radiative Transfer within Seagrass Canopies: Impact on Carbon Budgets and Light Requirements. *Proceedings of SPIE Ocean optics XIII*, 2963: 331-336

# Satellite Remote Sensing of Coral Reef Habitats Mapping in Shallow Waters at Banco Chinchorro Reefs, México: A Classification Approach

Ameris Ixchel Contreras-Silva[1], Alejandra A. López-Caloca[1],
F. Omar Tapia-Silva[1,2] and Sergio Cerdeira-Estrada[3]
*[1]Centro de Investigación en Geografía y Geomática
"Jorge L. Tamayo" A.C., CentroGeo
[2]Universidad Autónoma Metropolitana, Unidad Iztapalapa
[3]Comisión Nacional para el Conocimiento y
Uso de la Biodiversidad, CONABIO
Mexico*

## 1. Introduction

Interest in protecting nature has arisen in contemporary society as awareness has developed of the serious environmental crisis confronting us. One of the ecosystems most impacted is the coral reefs, which while offering a great wealth of habitats, diversity of species and limitless environmental services, have also been terribly damaged by anthropogenic causes. One example of this is the oil spill from petroleum platforms (in the recent case of the Gulf of Mexico). The effects of global warming—such as the increase in the incidence and intensity of hurricanes and drastic changes in ocean temperature—have caused dramatic damage, such as the bleaching and decrease of coral colonies. In light of this devastating situation, scientific studies are needed of coral reef communities and the negative effects they are undergoing.

The case study presented in this work takes place in the Chinchorro Bank coral reefs in Mexico. These are part of the great reef belt of the western Atlantic, with a biological richness that inherently provides environmental, economic and cultural services at the local scale as well as worldwide. Nevertheless, these services have been weakened for decades due to overexploitation, inducing imbalances and problems in the zone. Over recent decades, numerous biological communities that house constellations of species—whose natural evolutionary process dates back million of years (Primack et al., 1998)—have been alarmingly degraded. If this trend continues, the entire evolution that is sustained by the life of these communities will disappear in a relatively short period of time.

This study clearly demonstrates the application of state-of-art Remote Sensing (RS) in coral ecosystems. It includes an analysis based on the application of Iterative Self Organizing Data Analysis (ISODATA) as a classifier for generating classes of benthic ecosystems present in a coral reef system, using satellite images (Landsat 7-ETM+).

## 2. Use of remote sensing in coral reef ecosystems

The observation of the earth using remote sensors is a most complete method for monitoring the most significant natural risks (Xin et al., 2007). In general, RS has proven to be a powerful tool in the overall understanding of natural and anthropogenic phenomena. It is particularly appreciated as a non-invasive, non-destructive technique with global coverage. Thus, satellite, airborne and *in-situ* radiometry have become useful tools for tasks such as characterization, monitoring and the continuous prospecting of natural resources.

Research using RS has been strengthened in recent decades as a result of the growing concern worldwide for the preservation of coral reef systems as natural reservoirs. This has been observed to be an excellent method for analysis, which aids in the holistic study of this complex ecosystem. In order to develop an approach that helps to safeguard these ecosystems, it is necessary to understand the physical, chemical, biological and geological dynamics that occur therein (Brock et al., 2006). Andréfouët & Riegl (2004) refer to RS as a technology that is now virtually mandatory for research where spatial and temporal precision is required. RS has gone from being a tool with no application to coral reef systems to one that is *per se* indispensable. Andréfouët & Riegl (2004) discuss four reasons why this change has occurred:

- The proliferation of new sensors for acquiring direct and indirect data for monitoring coral reefs,
- The proliferation and improvement of analytical, statistical and empirical approaches,
- Recognition of global climate change due to anthropogenic human impacts that are lethal to coral reefs and
- Improved integration of technology for the conceptual design of coral reef research.

RS techniques offer an option for marine habitat mapping to determine not only the location and amount of different benthic habitats (Kirk, 1994) but also how these habitats are distributed and the degree of connectivity among them (Rivera et al., 2006). Nevertheless, the study of coral reefs using RS presents several important limitations. For example, intense cloud cover in optical images, optical similarities among spectral signatures of benthic communities, attenuation of the deep component (specific to each coral reef ecosystem) as well as the spatial and spectral resolution of remote sensors. In spite of these limitations, satellite sensors are highly useful for mapping the benthic bottom (Mumby et al., 1997), monitoring changes in its ecology (Krupa, 1999) and defining management strategies (Green et al., 1996).

### 2.1 Determination of ecological characteristics of coral reefs using remote sensors

Some of the characteristics of coral reefs that can be calculated using RS are temperature, wave height, sea level, turbidity, amount of chlorophyll and concentration of dissolved organic matter. In the case of atmospheric variables, it is possible to determine cloud cover, amount of seasonal rainfall, presence of contaminants and incidental solar energy (Andréfouët et al., 2003). All these factors directly and indirectly influence coral reefs and determine their health status (Andréfouët & Riegl 2004). In addition, it is possible to determine the different benthic ecosystems present in the coral reefs, such as seagrass, type of bottom, algae communities and different types of coral. If the reef is near a tourist or vacation area, anthropogenic impacts can be determined by calculating the growth of the urban stain, vegetation coverage, the structure of the hydrographic basins, etc. Intrinsic conditions of coral reefs can be described, which are largely defined by the inflows and outflow and their transport of sediments and export of dissolved organic matter. This enables us to understand the patterns involved in coral whitening, among other events (Brock et al., 2006).

The coral reefs—located in relatively clear water—allow us to use passive optic sensors (Benfield et al., 2007). The more common satellite sensors that have been used to study this are SPOT, Landsat TM and ETM+ (Andréfouët & Riegl 2004; Benfield et al., 2007; Mumby 2006; Mumby et al., 2004; Mumby and Harborne 1998). Studies previously conducted (Green, 2000; Mumby et al., 1999) have observed that Landsat and SPOT images are suitable for mapping corals, sands, and seagrass, depending on their resolution. Nevertheless, it is important to note that various types of habitats can be represented in one Landsat image pixel (or others with less spatial resolution), which may limit classification abilities (Benfield et al., 2007). Previous studies conducted (Green, 2000; Mumby et al., 1999) have observed that according to the resolution of Landsat images, they are suitable for mapping sea corals, sands and seagrass. Based on this assumption, the data obtained from Landsat and SPOT are adequate for simple complexity mapping (3-6 classes, such as seagrass, sand, dead corals and some species of corals) but for more complex targets (7-13 classes) they are limited by their spatial and spectral resolution. (Mumby, 1997; Andréfouët et al., 2003; Capolsini et al., 2003). To a lesser extent, SeaWiFS (sea-viewing wide field of view sensors) have also been used, as well as IKONOS with higher spatial resolution, LIDAR and SONAR, among others (Andréfouët & Riegl 2004; Andréfouët et al., 2003; Brock et al., 2006; Elvidge et al., 2004; Liceaga-Correa & Euan-Avila, 2002; Hsu et al., 2008; Lesser and Mobley, 2007). It is important to note that analytical methods as well as spatial modeling, statistics and empirical methods at different scales and for different applications have been used in direct relation to ecological processes of reefs (Andréfouët & Riegl 2004). The use of airborne remote sensors, such as CASI (Compact Airborne Spectrographic Imager) with a high spectral or hyperspectral resolution, has gradually been increasing in this type of studies, to the extent that the specialists mention that mapping reefs using air or satellite sensors have proven to be more effective than fieldwork (Mumby, 1999). Nevertheless, field measurements cannot be discarded, since they provide us with the basis for corroborating the information obtained from satellite images. In addition, images from satellite sensors provide the opportunity to conduct multi-temporal monitoring (Helge et al., 2005) in order to identify the status of an ecosystem and predict possible future changes.

According to the above, it can be stated that studies applying RS in coastal ecosystems and, specifically, in coral reef ecosystems provide information and knowledge that can successfully be applied to define management strategies for these important ecosystems, as well as to design viable alternatives for their conservation.

## 3. Spectral reflectance of coral

To make observations, we move vertically and gradually from the coral surface to the water surface, measuring the changes in the quantity of light in the water column that falls directly on the coral. The quantity of light present obviously affects the amount that is reflected by the coral, and is therefore a crucially important parameter for mapping it.

Spectral reflectance (ρ) is a key parameter for conducting studies of coral reefs using RS (Hochberg et al., 2004). Two factors clearly and concisely explain this. First, ρ represents the boundary of radiative transference in the water surface optics. Therefore, taking into account ρ can resolve the problem of inverse radiative transference presented by passive remote sensors when applied in this field. Second, ρ is the function that denotes the object, the composition of the material and its structure. Therefore, it serves as a bridge between the optics of the object and the shape of the sea bottom (Hochberg et al., 2004).

In the process of classifying images and generating thematic maps, large differences have been noted in spectral reflectance among the coral reefs' benthic communities (Brock et al., 2006). Variability in the vertical relief, or rugosity, is a significant aspect of the complexity of a habitat, a factor that both reflects and governs the spatial distribution and density of many reef organisms (McCormick 1994). These factors, which respond to these evaluations, vary according to the differences among sediments, the presence of different algae species and the coverage of atypical algae in surface water in some reef zones. Thus, Hochberg et al. (2004) mention the importance of creating a specific approach using RS to study the surface water mass presented by atypical algae, since it has been shown that the mere presence of these organisms indicates classes that are spectrally distinct from other reef communities, even when they represent the same species.

Differences among the spectral signatures of corals provide a high likelihood of satisfactorily delineating and defining their different features in a satellite image. The problem with the above process is that the ρ of the corals is a function of pigmentation, structure, the orientation of their branches and their internal characteristics (Newman et al., 2006). In addition, though the interactions between light and the atmosphere are well-studied, the challenge is to establish controls for the effects of the water column in which the coral is found that influence these factors. Taking into account the curvatures in order to correct the acquired data provides more valuable information about the conditions and health of the living communities sheltered by the coral. Newman et al. (2006) indicate that two categories have been defined by recent studies which were conducted to measure in situ the spectral signatures of the coral environment:

i.    The spectral signatures are examined according to the variation in the pigment density, which characterizes the sensorial color of the different coral species (Newman et al., 2006). Some studies have analyzed the contribution of color to the measurement of radiance (R), in particular, by comparisons with unpigmented coral structures. These observations resulted in the spectrum of coral whitening and structures saturated with zooxanthellae (Newman et al., 2006), which provide a measure of the health status of the complex reef system. Color has been used as a comparison measurement among three coral species, five algae species and three benthic communities (Hochberg and Atkinson, 2000), and as a means to differentiate between dead coral in different stages and algae colonization (Clark et al., 2000).

ii.   Spectral signatures were examined according to morphological characteristics (Newman et al., 2006).
      Corals exhibit distinct and complex structural morphologies, partially due to environmental conditions such as light availability, water motion and suspended sediment (Joyce & Phinn, 2002). Reflectance values measured over varying angles and azimuths were examined to determine the bidirectional reflectance distribution

function of coral species and the inter-species variation between rounded and branching types (Joyce and Phinn, 2002; Newman et al., 2006).

## 4. Mapping coral reefs using remote sensors

The worldwide importance of coral reefs in light of current threats has generated interest in developing methods to study this type of ecosystems at global scales (Kuhn 2006). The use of remote sensing to map underwater habitats is increasing substantially. This enables using the derived information to determine the status of these natural resources as a basis for planning, management, monitoring, conservation and evaluating their potential.

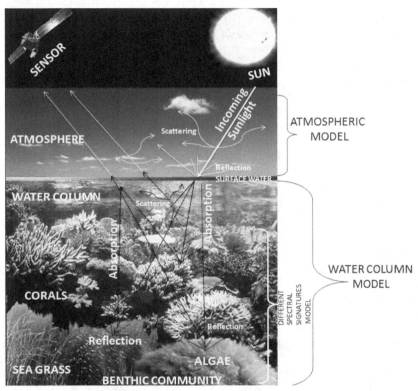

Fig. 1. Components of Remote Sensing in mapping coral reefs.

As was mentioned previously, high resolution spectral sensors exist that have elements that enable specific analysis with an excellent capacity for modelling environmental and structural variables in the coral reefs (Holden and LeDrew, 1998). The data produced by this type of sensors provide products that can be combined with models to photosynthetically calculate the radiation available through the photic zone and the surface of benthic substrates. Established models for calculating incident solar radiation are developed and evaluated based on routine satellite and meteorological observations (Brock et al., 2006). The spectral differences among corals, seagrass and algae are nearly imperceptible and not easy to detect with the three bands (blue, green and red) of the sensors that can penetrate the water column (Holden and LeDrew 1998; Hedley and Mumby 2002; Karpouzli et al., 2004). This is why RS studies applied to the mapping of submerged benthic ecosystems requires the generation of new processing methodologies. In addition, coral habitats present a heterogeneity that is inherent of their complexity, and therefore the task of discerning among the different spectral signatures is more complicated. That is, the pre-processing of images applied to this type of environments should not only incorporate the elimination of noise in the atmospheric and batimetric portions, but should also take into account the components of the water column, as shown in Figure 1.

## 5. Pre-processing of satellite images

All satellite images must undergo an initial processing of crude data to correct radiometric and geometric distortions of the image and eliminate noise. It must be taken into account that the energy captured by the sensor goes through a series of interactions with the atmosphere before reaching the sensor. As a result, the radiance registered by the sensor is not an exact representation of the actual radiance emitted by the covering. This means that the image acquired in a numerical form presents a series of anomalies with respect to the real scene being detected. These anomalies are located in the pixels and digital levels of the pixels that make up the data matrix. The purpose of correction operations is to minimize these alterations. The corrections are made during pre-processing operations, since they are carried out before performing the procedures to extract quantitative information. The product obtained is a corrected image that is as close as possible, geometrically and radiometrically, to the true radiant energy and spatial characteristics of the study area at the time the data are collected. Atmospheric correction is a process used to reduce or eliminate the effects of the atmosphere and allow for more precisely seeing the reflectance values of the surface being studied or analyzed.

Nevertheless, when attempting to map or derive quantitative information from subaquatic habitats, the depth of the water significantly affects the measurements taken by remote sensors, making it possible to generate confusion about spectral signatures. Therefore, atmospheric and geometric corrections are not sufficient when the objective is to extract features of the covering of the bottom of the water. That could be considered a characteristic and, in some cases, a limitation of passive sensors in remote sensor applications in marine environments. Thus, in this type of studies, a water column correction is performed to improve reliability when analyzing the results of the image and to eliminate the noise resulting from the variation in the ground's reflectance (Holden 2002; Holden and LeDrew, 1998; Mumby, 1998).

### 5.1 Correction of remotely sensed imagery

### 5.1.1 Radiometric correction

The radiance from the sensor (L) is calculated as:

$$L = c0 + c1*ND \tag{1}$$

Where c0 and c1 are the offset and gain, respectively, of the radiometric calibration and ND is the digital number recorded in a particular spectral band. The process of obtaining L is called radiometric correction.

The total signal captured by the sensor consists of three parts: atmospheric scattering of radiation, radiation reflected by the pixel and radiation reflected by the vicinity of the pixel and scattered in different (adjacent) directions.

### 5.1.2 Atmospheric correction

The atmospheric conditions (water vapor, aerosols and visibility) in a scene can be calculated using algorithms that are performed using a database based on atmospheric functions. The surface spectral reflectance of an interaction target in a scene can thereby be seen as a function of the atmospheric parameters. ¶(6pt)

### 5.1.3 Geometric correction

The geometric correction consists of distinguishing the other types of radiation and only considering that which is reflected by the pixel. The objective is to remove geometric distortion; that is, to locate each pixel in its corresponding planimetric position. This enables associating the information obtained from a satellite image with thematic information from other sources.

### 5.2 Water column correction

The coral reefs generally develop in transparent or clear water, which facilitates study and analysis with passive optic, multispectral or hyperspectral sensors (Mumby et al., 1999). When light penetrates the water column, its intensity exponentially decreases as the depth increases. This process is known as attenuation, and it has an important effect on data obtained by remote sensors in aquatic environments (Green, 2000). The attenuation process is shown in Figure 2.

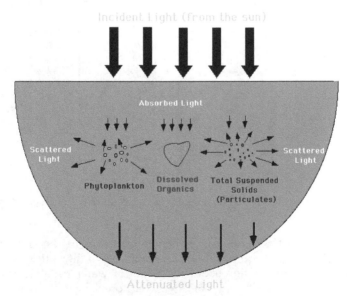

Fig. 2. Processes of light attenuation in the water column (SERC, 2011).

There are two reasons for this phenomenon:

- Absorption: light energy is converted into another type of energy, generally heat or chemical energy. This absorption is produced by the algae, which utilize the light as a source of energy, by suspended organic and inorganic particulate matter (OPM and IPM), dissolved inorganic compounds and the water itself.
- Scattering: This phenomenon results from the collision of light rays and suspended particles, causing multiple reflections. The more turbid the water (more suspended particles) the greater the scattering effect, making it difficult for light to penetrate.

The attenuation varies according to the wavelength of the electromagnetic radiation (EMR). For example, in the region of visible light, the red portion of the spectrum attenuates more quickly than the short wavelength, such as blue.

Figure 3 shows, for 4 spectral bands (blue, green, red and infrared), how the spectrum in a particular habitat (seagrass or macroalgae) can change as the depth increases. The spectral radiance registered by a sensor is dependent on the reflectance of the substrates and the depth. As the depth increases, the possibility to discriminate spectrums or spectral signatures of the habitats decreases. In practice, the spectrum of sand at a depth of 2 meters is very different than that at 20 meters. According to Mumby and Edwards (2000), the spectral signature of sand at 20 meters could be similar to that of seagrass at 3 meters. All these factors influence the signal and can create a good deal of confusion when using visual inspection or spectral classification to classify these habitats. Therefore, the influence of the variability in depth must be eliminated, which is known as water column correction or depth correction (Mumby and Edwards 2000).

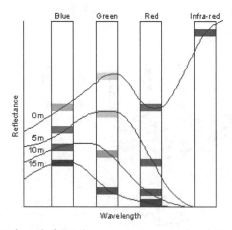

Fig. 3. Spectral differences for a habitat (seagrass or macroalgae) at different depths (Mumby and Edwards, 2000).

A variety of models exist that can be used to compensate for the effect of the water column. Nevertheless, many require optical measurements of the optical properties of the water itself, as well as information about the depth of water per pixel (Gordon, 1978; Philpot, 1989; Mobley et al., 1993; Lee et al., 1999; Maritorena et al., 1994; Maritorena 1996; Lee et al., 1999). Thus, the method proposed by Lyzenga (1981) is applied, which has been used and described by other authors, such as Mumby et al., 1997, 1998, Mumby and Edwards 2002, Andréfouët et al., 2003, etc. This approach has the advantage of taking into account the majority of the spectral information and not requiring data for the components of the water surrounding the reef. Instead of deriving the spectra of the different types of sea bottoms and water properties, this method transforms the spectral values into "depth-invariant indices." The primary limitation of this method, among others, is that it must be applied to clear water (i.e. type 1 or type 2); the study area meets this requirement.

To eliminate the influence of depth on sea bottom reflectance, the following need to be taken into account: the identification of the characteristics of attenuation of the water column and having digital models of the depth; although these are not very common, particularly for coral reef systems (Clark et al., 2000). This work used a bathymetric model provided by SEMAR (2008) that makes possible a good deal of reliability and precision to the measurements.

The procedure is divided into various steps:

1. Elimination of the atmospheric scattering and the external reflection from the water surface (atmospheric correction). This can be carried out using a variety of methods, such as dark pixel subtraction (Maritorena, 1996) and ATCOR (Richter, 1996, 1998).
2. Selection of pixel samples with the same substrate and different depths.
3. Selection of a spectral band pair, with good penetration of the water column (that is, bands found in the visible light spectrum — Landsat TM and ETM+ 1/2, 2/3 and 1/3).
4. Linearization of the relationship between depth and radiance, Xi = ln (Li), where Xi is the transformed radiance of the pixel in band i (band 1) and Li is the radiance of the pixel in band j (band 2). When the intensity of the light (radiance) is transformed using the natural logarithm (ln), this relationship becomes linear with the depth. Therefore, the transformed radiance values will decrease linearly as depth increases:

$$X_i = Ln(L_i) \tag{2}$$

5. Determination of the attenuation coefficient (quotient) using a biplot of the transformed radiance of the 2 bands (Li and Lj). The biplot contains data for one type of uniform bottom (sand) and variable depth. It is created using the following equations:

$$K_i/K_j = a + \sqrt{(a^2 + 1)} \tag{3}$$

$$a = \frac{\sigma_{jj} - \sigma_{ii}}{2\sigma_{ij}} \quad \text{and} \quad \sigma_{ij} = \overline{X_i X_j} - \overline{X_i}\,\overline{X_j} \tag{4}$$

where $\sigma_{ii}$ is the variance in band i and a is the covariance between bands i and j.

6. Lastly, the depth-invariant index is generated using the equation by Lyzenga (1981):

$$IIP_{ij} = \ln(L_i) - \left[ \left( \frac{k_i}{k_j} \right) \ln \right] (L_j) \tag{5}$$

The result of this operation generates a new band—the image with water column correction for a band pair (depth-invariant index). Since the values of this band are whole numbers with decimals and can be negative, in order to visualize them they need to be converted into an 8-bit format, that is, gray values between 0 and 255. To this end, minimum and maximum values for the resulting image must be found and linearly distributed between the values 1 and 255 (0 is not included because it is assigned to the masked surface area). The depth-invariant index is essential when the objective of the study is to extract spectral data for submerged aquatic environments.

## 6. Review of classification methodologies

The classification of a satellite image consists of assigning a group of pixels to specific thematic classes based on their spectral properties. The spatial classification of underwater coastal ecosystems is one of the most complex processes in thematic cartography using satellite images. As previously mentioned, this can be attributed primarily to the influence of the atmosphere and the ocean water column, through which electromagnetic radiation passes. In addition, it is worth mentioning that these ecosystems undergo constant variation, especially after significant events such as strong hurricanes. Nevertheless, different authors (Mumby et al., 1997; Andréfouët & Payri 2000; Mumby and Edwards 2002; Andréfouët et al., 2003; Pahlevan et al., 2006; Call et al., 2003, etc.) have been using remote sensing to develop different classification methods for these ecosystems and, in particular, for coral reefs.

The maximum likelihood classifier is the most common method, and has been used by authors such as Mumby et al. (1997), Andréfouët et al. (2000), Mumby and Edwards (2002), Andréfouët et al. (2003), Pahlevan et al. (2006), and Benfield et al. (2007). Its primary advantage is that it offers a greater margin for accounting for the variations in classes through the use of statistical analysis of data, such as the mean, variance and covariance. The results of the method can be improved with the incorporation of additional spatial information during the post-classification process, since this helps to spectrally separate the classes that had been mixed.

Another method also used by Mumby et al. (1997) is agglomerative hierarchical classification with group-average sorting. An alternative proposal is object-oriented classification, which consists of two steps, segmentation and classification. Segmentation creates image-objects and is used to build blocks for further classifications based on fuzzy logic. Another method that has been used is ISODATA (iterative self-organizing data analysis), which uses a combination of Euclidian squared distance and the reclassification of the centroid (Call et al., 2003). In this study, ISODATA was used to perform the classification.

### 6.1 ISODATA (Iterative Self Organizing Data Analysis)

ISODATA is an unsupervised classification method as well as a way to group pixels, and uses the minimum spectral distance formula. It begins with groups that have arbitrary means and each time the pixels in each of the iterations are regrouped and the means of the groups change. The new means are then used for the next iterations.

The algorithm for obtaining the classification is based on the following parameters:

Satellite Remote Sensing of Coral Reef Habitats Mapping in Shallow Waters at Banco Chinchorro Reefs,
México: A Classification Approach

25

a. The user decides on the number N of clusters to be used. For the first calculation, it is recommended to use a high number, which is then reduced by interpreting the image.
b. A set of N clusters in the space between the bands is selected. The initial location is in the zones with the highest reflectance.
c. The pixels are assigned to the closest cluster.
d. The clusters are associated, dispersed or eliminated depending on the maximum distance of the class or the minimum number of pixels in a class.
e. The grouping of pixels in the image is repeated until the maximum number of iterations has been reached, or a maximum percentage of pixels are left unchanged after two iterations. Both parameters can be specified

## 7. Case study

The Chinchorro Biosphere Reserve (Fig. 4) is located in the open Caribbean Sea, 30.8 km east of the coastal city of Mahahual, which is the closest continental point. The coral reef of Chinchorro Bank, Mexico, is part of the great reef belt in the western Atlantic, the second largest in the world, and is the biggest oceanic reef in Mexico. With a reef lagoon area of 864 km2, it is considered a pseudo-atoll or reef platform (Camarena, 2003). Chinchorro Bank is a reef complex that contains an extensive coral formation with a vast wealth and diversity of species and high ecological, social and cultural value. It inherently provides certain services, including the protection of the coast from battering by storms and hurricanes. The area has been exploited by fishing and tourist-related scuba diving over the past decades. The Chinchorro Bank supports pristine reefs, coral patches, extensive areas of seagrass, microalgae beds and sand beds. The reserve's ecosystems are marked by mangroves and reef zones. The composition of the taxocenosis of coral is known to contain hexacorals,

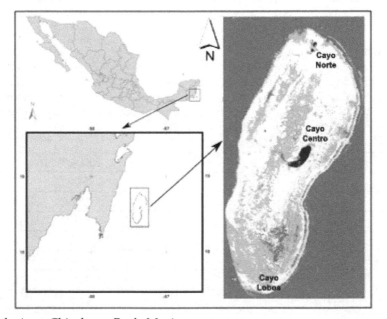

Fig. 4. Study Area: Chinchorro Bank, Mexico.

octocorals and hydrozoas and a reported 95 different species (Camarena, 2003). The diversity of the fauna in the Chinchorro Bank is very high and includes several phyla, families, genres and species, with at least 145 macro invertebrate and 211 vertebrate species, in addition to the corals (Bezaury et al., 1997).

The biogeographic region of Chinchorro Bank is delimited on the north by the Caribbean Province which extends along Central and South America. This province begins in Cabo Rojo, in southern Tampico, and extends into eastern Venezuela and the northern Orinoco delta. The land biota is greatly similar to that of the continent and is therefore considered to be part of the Yucateca Province. It is located in the Mexican Caribbean, across from the southeastern coast of the state of Quintana Roo, between the 18°47'-18°23' N and 87°14'-87°27' W parallels. It is 30.8 km from the continent and separated from it by a wide canal 1000 m deep. The shape of Chinchorro Bank is elliptic, with a reef lagoon that includes a sandy bank 46 m long (north-south) and 18 km wide (east-west) at its broadest part. The total area is 144360 ha. The periphery of the bank is bordered by active coral growth on the eastern (windward) margin, which forms a coral reef, or breaker, while along the western margin (leeward) the breaker disappears and the coral growth is semicontinuous and diffuse (Camarena, 2003). There are four emerged zones within the bank—known as "Cayo Norte" (two islands), "Cayo Centro" and "Cayo Lobos" —whose ecological value is very high because of their diverse species of land and water flora and fauna (Camarena, 2003).

## 8. Information resources

The geospatial database used in this study includes a Landsat 7-ETM+ image (Table 1), bathymetric information and in situ data for sand (Figure 5). The digital data were projected to UTM (Universal Transverse Mercator) zone 16 north with WGS-84 datum. ERDAS, GEOMATIC 10.2 and ArcMap 9.3 were used to process the data.

The importance of choosing the type of image with which to work is well-known, particularly because the users will need to make sure to use images that are suitable to the purpose of the study. The nature of a platform-sensor system determines the characteristics of the image's data (Green, 2000). The Landsat 7-ETM+ (Table 1) image obtained had no cloud cover. It is worth noting that this type of images provides adequate coverage of the area for regional and temporal monitoring studies.

| Date | 2000-03-29 |
|---|---|
| Scan time | 16:03:05 |
| Path/Row | 18/47 |
| Spatial resolution (m) | 30 |
| Spectral bands used | 3 |
| Spectral range (μm) | 0.5-0.69 |
| Azimuth | 116.29 |
| Solar angle | 59.43 |

Table 1. Characteristics of the Landsat 7-ETM+ image used

Satellite Remote Sensing of Coral Reef Habitats Mapping in Shallow Waters at Banco Chinchorro Reefs,
México: A Classification Approach

27

It is also very important to note that bathymetry is one of the most relevant factors in the dynamic ecology of coral reefs. Numerous reef studies show that coral species diversity tends to increase as a function of depth, reaching its maximum between 20–30 m and diminishing with greater depth (Huston, 1985). This depth effect results in a marked zonation of the reef community (Aguilar-Perera and Aguilar-Dávila, 1993). While the upper depth limits of corals are controlled by various physical and biological factors, their maximum depth depends largely upon light availability (García-Ureña, 2004). The bathymetric soundings for Chinchorro Bank used by this study were done in 2008 by the Mexican Navy (SEMAR, 2008). The depth of the interior of the bank varies. The northern portion is shallower, between 1 and 2 m, the depth of the central portion ranges between 3 and 7 m, and the southern portion is deepest, varying between 8 and 15 m (SEMAR, 2008). There are 4 emerged zones within the bank, known as keys, which have high ecological value because of their diverse species of flora and aquatic and land fauna (Camarena, 2003). Figure 5b shows bathymetry data for the Chinchorro Bank, where the depths of the zone can be seen.

In situ sampling data were provided by SEMAR. Data from Carricart-Ganivet et al. (2002) were also used. Based on these data, 4 of the most representative classes were determined: 1) coral mass, 2) coral patches, 3) seagrass and algae and 4) sand. The ocean and keys, or emerged areas, are not part of the classification criteria, though they are also represented. Unfortunately, the databases for the in situ sampling have disadvantages—such as mixing classes in the same point and lack of definition of the benthic bottom, among others—that prevent their being used for validation purposes. Only data for sand provided by SEMAR do not present these disadvantages and could be used for water column correction, as explained further below.

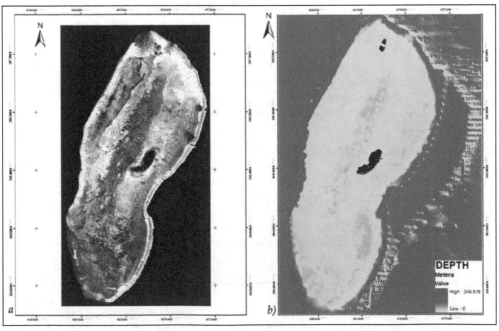

Fig. 5. Information resources. a) Landsat 7-ETM+ image and b) depth of the Chinchorro Bank

## 9. Results and discussion

### 9.1 Image processing

A Landsat 7-ETM+ image from March 29, 2000 was processed. Before conducting the quantitative analysis of the data, a post-calibration was performed of the constant gain and offset to convert the image ND to spectral radiance. The spectral radiance was also corrected for atmospheric effects to obtain the surface reflectance values. A geometric correction was not performed because the level of the processing of the Landsat images includes this correction. Only 3 of the 8 bands contained by Landsat were used (blue, green and red). The depth correction was developed with the Lyzenga (1981) method, which has been used and described by other authors (Mumby et al., 1997, 1998; Mumby and Edwards 2002; Andréfouët et al., 2003).

### 9.2 Water column correction

Lyzenga (1981) shows that when drawing a scatterplot of 2 of the logarithmically transformed bands in the visible spectrum (one on each axis), the pixels for the same type of bottom (i.e. sand at different depths) follow a linear trend. Repeating this process for different types of bottoms produces a series of parallel lines and the intersection of those lines generate a unique depth-invariant index which is independent of the type of bottom; all the pixels for a particular bottom have the same value as the index regardless of the depth at which they are found (Andréfouët et al., 2003). A group of pixels representative of the depth of the water column was selected for this study, therefore pixels very close to the surface (< 1m) were eliminated. Sand was the only substrate used since it is the most homogenous bottom in coral environments, and is the one most used by various authors (Mumby and Edwards 2002; Lyzenga, 1981) and the most easily recognizable for interpretation purposes. For the specific case of the Chinchorro Bank, 100 points of sand between 1 and 10m of depth were used to determine the attenuation coefficient (quotient) for the band pair ½, 99 points were used for bands 1/3 and 96 for bands 2/3. The data for point radiance to a type of bottom were extracted from the image and transferred to a spreadsheet. Figure 6a shows the graphic spectral radiance of bands 1 and 2 (atmospherically corrected) with respect to the depth for one specific type of bottom (sand) and variable depth.

Figure 6b shows the linearization of the exponential attenuation of the radiance for bands 1 and 2 using natural logarithms, since in practice it is virtually impossible for the points to adhere to a perfect line given the natural heterogeneity of the different types of bottoms, variations in the water quality, surface roughness of the water, etc. Figure 6c shows the biplot of bands 1 and 2 for a single substrate (sand) at different depths. To this end, the variance of band 1 and the covariance of bands 1 and 2 are evaluated (Table 2 and 3). Table 3 shows the different values for obtaining the attenuation coefficient, according to spectral band.

|                          | Band 1 | Band 2 | Band 3 |
|--------------------------|--------|--------|--------|
| Variance ($\sigma_{ii}$) | 0.2628 | 0.6334 | 0.2761 |

Table 2. Variance of the radiance of each band

Satellite Remote Sensing of Coral Reef Habitats Mapping in Shallow Waters at Banco Chinchorro Reefs, México: A Classification Approach

29

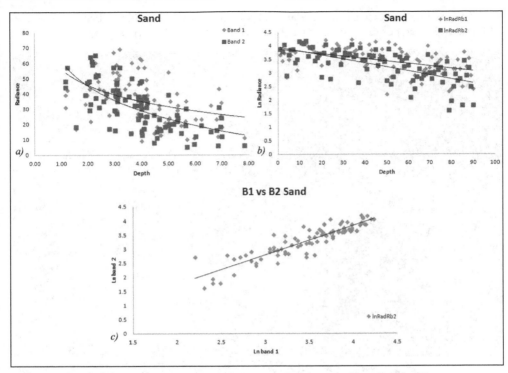

Fig. 6. Steps for water column correction: (a) spectral radiance of bands 1 and 2 (atmospherically corrected), (b) exponential decay of the radiance for bands 1 and 2 using natural logarithms and (c) biplot of bands 1 and 2 for a single bottom (sand) at different depths.

|  | Ratio 1/2 | Ratio 1/3 | Ratio 2/3 |
|---|---|---|---|
| Covariance ($\sigma_{ij}$) | 0.3200 | 0.1178 | 0.2327 |
| aij | -0.0593 | -0.0031 | 0.0184 |
| ki/kj | 0.94 | 0.99 | 1.00 |

Table 3. Calculation of ratio of attenuation coefficients

Figure 6c shows the biplot of the logarithmically transformed bands 1 and 2, representing the attenuation coefficient (ki/kj) for bands 1 and 2. It is important to mention that if different types of bottoms are represented in a biplot, they would theoretically represent a line with a similar behavior, varying in position only due to differences in spectral reflectance. The gradient of the line would be identical since ki/kj does not depend on the type of bottom. The intersection of the line with the y-axis represents the depth-invariant index, since each type of bottom has a unique y-intersect regardless of depth. Each pixel is assigned an index depending on the type of bottom, which is obtained using the natural logarithm transformation for each band and the connection of the coordinate to the origin of the y-axis through gradient line ki/kj. The pixels are thus classified for different types of bottoms.

As mentioned before, the depth-invariant index is generated according to band pairs—1/2, 1/3 and 2/3, correponding to bands 1 (blue), 2 (green) and 3 (red) (Figure 7). The image

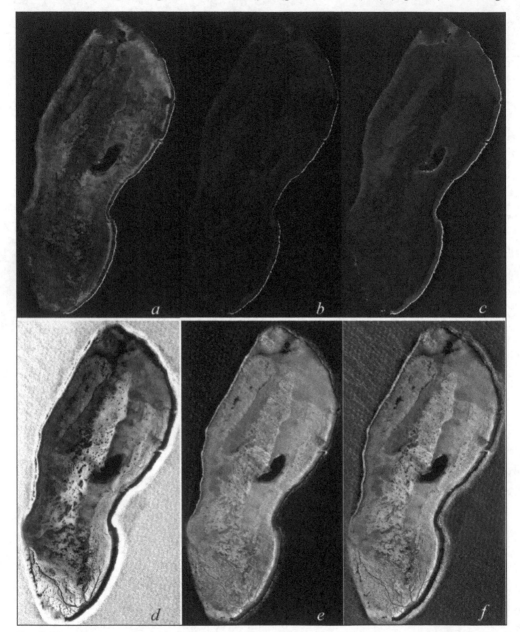

Fig. 7. Visualization of the Landsat 7-ETM+ image before and after water column correction. a) image of band 1 (blue, 450-520 nm), b) band 2 (green, 530-610 nm), c) band 3 (red, 630-655 nm), d) depth-invariant index combination of bands 1/2, e) 2/3 and f) 1/3.

Satellite Remote Sensing of Coral Reef Habitats Mapping in Shallow Waters at Banco Chinchorro Reefs, México: A Classification Approach

31

resulting from the depth-invariant index was significantly different than the image without correction, since it reveals more details of the structures of the benthic bottom, especially in zones with greater depths.

## 9.3 ISODATA classification

As an initial approach to the classification of submerged benthic ecosystems in the Chinchorro Bank, ISODATA was used as a classification method, since not much needs to be known about that data beforehand. A little user effort is required to identify spectral clusters in data. The results of the benthic classification in the Chinchorro Bank were visually evaluated according to the quality of the segmentation using the classification by Aguilar-Perera & Aguilar-Dávila (1993), and with bathymetric data that greatly determine the ecology of the corals, as explained next.

Figure 8a shows the Landsat image with atmospheric correction for the RGB (1,2,3) combination and Figure 8b shows the image resulting from the depth-invariant index by bottom type. At the bottom of the figure, two images classified using ISODATA are included, both with the same type and number of classes. Figure 8c presents the classification performed without water column correction; that is, using the image from 8a as input. Figure 8d includes the classification performed based on the depth-invariant index (shown in 8b); that is, taking into account water column correction. To identify the categories resulting from the ISODATA process, benthic bottoms in the Chinchorro Bank as defined by Aguilar-Perera & Aguilar Dávila (1993) were used as a basis. It can be seen (8c) that the classification without water column correction produced a substantial mix of classes throughout the image, unlike the classification obtained by applying water column correction (8d). According to authors such as Aguilar-Perera & Aguilar Dávila (1993), Chávez and Hidalgo (1984) and Jordán (1979), the periphery of the Chinchorro Bank is surrounded by abundant coral growth on the eastern margin. A barrier reef is thereby formed that disappears along the western margin where the coral growth is semi-continuous and diffuse. This spatial distribution of the corals can be clearly seen in the results of the classification with water column correction (Figure 8d), unlike classification without correction (Figure 8c).

One known ecological characteristic of reef systems is that the zonation of the reef bottom and its ecological dynamics are strongly influenced by the depth (Huston, 1985; Loya, 1972; Gonzáles et al., 2003). The seagrasses constitute a type of benthic bottom normally present in shallower zones. These observations and the use of bathymetry enable corroboration of the validity of the spatial distribution of seagrasses obtained by classification with water column correction. The shallower zones are located in the northern (1-2m) and central (3 and 4 m) portions; these two zones best correspond to the zone with seagrass generated in the image shown in 8d, as opposed to the image in 8c where it can be seen that the seagrass class is distributed throughout the bank. In addition, 8c shows a mix between seagrass and corals, a result that is not justifiable since the corals normally develop at depths between 5 and 30m. Using the depth criterion again in order to define the zonation, it is possible to state that the classification with water column correction produces good results for identifying coral patches, since they are found at depths between 7 and 12 m, as can be seen in Figure 8d. As a general observation, we can state that the results of the classification with water column correction generate data that are consistent with the theory regarding the

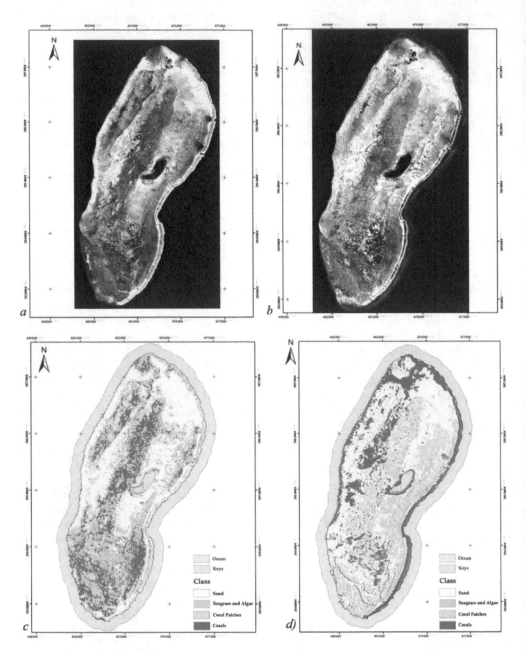

Fig. 8. a) Landsat 7-ETM+ image, RGB (1, 2, 3), b) image resulting from the depth-invariant index by bottom type using bands 1 and 2, and classification of the benthic bottom in the Chinchorro Bank using ISODATA, c)without water column correction and d) with water column correction.

Satellite Remote Sensing of Coral Reef Habitats Mapping in Shallow Waters at Banco Chinchorro Reefs, México: A Classification Approach

33

influence of depth in defining the zonation of benthic bottoms, as well as observation of other authors regarding the spatial distribution of sea-bottoms.

Figure 9 shows a close-up to facilitate the visual analysis of the differences between the classes obtained using ISODATA, implemented with and without water column correction.

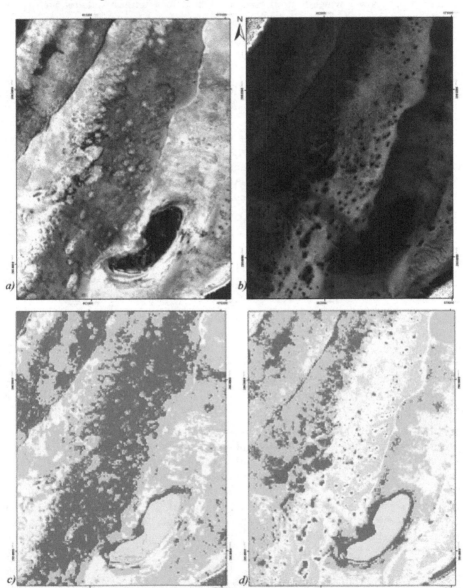

Fig. 9. Comparison among a) Landsat 7-ETM+ image, RGB (1, 2, 3), b) depth-invariant index by bottom type for bands 1/2, c) ISODATA without water column correction and c) ISODATA with water column correction.

In this figure, it can be seen that thanks to the water column correction, the classes are better defined, with mixing among them—caused by interference by the depth of the water column—avoided to whatever extent possible. The ISODATA algorithm more accurately selects and groups clusters, eliminating this problem. This visualization again confirms the advantage of performing water column corrections to obtain better results for the processes to classify benthic bottoms.

## 10. Conclusions

The study shows that the application of new remote sensing methods is crucial to the pre-processing of images in order to identify submerged aquatic ecosystems. This is because when quantitative information is mapped or derived from satellite images of aquatic environments, the depth of the water causes spectral confusion and therefore significantly affects the measurements of submerged habitats. Water column correction minimizes this effect, which enables distinguishing the classes of benthic ecosystems present in the Chinchorro Bank and demonstrates improvement especially in zones representing more variation in depth. Thus, water column correction is an indispensible pre-processing method in the cartography of submerged aquatic ecosystems.

The water column correction method used in this study uses the majority of the spectral information while disregarding the characteristics of the water surrounding the reef, such that the spectral values are transformed from a band pair into a depth-invariant index. This should be applied in relatively clear water (type 1 or type 2), as is the case of the Chinchorro Bank. Using this process, the attenuation effect of the water column was minimized, which is one of the primary problems with the segmentation of images of submerged ecosystems.

Traditional, unsupervised classification methods, such as ISODATA, have difficulty detecting subclasses, that is, this type of classifier makes it complicated to detect pixels between very close classes with distributions that share an overlapping zone. When classifying benthonic habitats in the Chinchorro Bank, it was possible to observe that the classes with less concentration of pixels were masked by those with greater amounts. This may be because standard methods, such as ISODATA, use moving mass center techniques to locate the classes and, thus, what are called subclasses become undetectable.

In general, the data from remote sensors are used for mapping reef habitats. Although the classification presented here was quite general—only 4 classes were determined—the results show that the Landsat 7-ETM+ images are able to identify different classes in submerged benthonic environments. Although the classification resulted in visually optimal results, the need to incorporate statistical validation of the data is important, so as to determine the accuracy of the classification performed in comparison to the reality; this was not possible for this study because an adequate database of in situ sampling was not available. Nevertheless, because of the visual comparison with classes identified by studies such as those by Aguilar-Perera & Aguilar Dávila (1993), Chávez and Hidalgo (1984) and Jordán (1979) and the consistency with the theory of the zonation of benthic bottoms based on depth, it can be concluded that the classifications obtained by ISODATA successfully determined the majority of the benthonic cases defined in this study of the Chinchorro Bank.

Coral reefs are being threatened worldwide by a combination of natural and anthropogenic impacts. Although the natural impacts are intense, there are intermediate time lapses that can contribute to maintaining biodiversity. On the other hand, the human impacts—which may seem to be less intense because they are not as perceivable to the eye—are chronic and can unleash a chain of negative effects. This sequence of negative effects normally does not give ecosystems the opportunity to recover and maintain their characteristic function and structure.

The search for new methodologies to process satellite images is indispensable to identifying the current trend in the degradation of marine habitats; methodologies that generate new and improved classifications that are highly reliable and with a level of detail that is adequate for mapping these ecosystems. Through this type of study, it is possible to organize, relate and manage information from satellite images in order to propose agreed-upon strategies to conserve natural resources, as part of comprehensive environmental policies to properly solve the problems. Thus, these data can be used as a basis to plan the monitoring of reefs in order to create scientific methods to generate knowledge and environmental awareness in the society and to contribute to the mitigation of the loss of reefs due to impacts from current global warming and other anthropogenic and global changes.

## 11. Acknowledgments

The authors would like to thank the Mexican Navy (SEMAR), Deputy Department of Oceanography, Hydrography and Meteorology (Dirección General Adjunta de Oceanografía, Hidrografía y Meteorología) for the information provided regarding bathymetry and the field sampling of sand data. We also thank Dr. Juan Pablo Carricart Ganivet and Janneth Padilla Saldívar for the information and geographic basis from the Comprehensive Management of the Chinchorro Bank: Geographic survey and geomorphologic characterization of the reef.

## 12. References

Aguilar-Perera A. & Aguilar-Dávila W., (1993). Banco Chinchorro: Arrecife Coralino en el Caribe. Pp. 807-816. In: Biodiversidad Marina y Costera de México, Salazar-Vallejo, S.I. & González, N.E. pp. 1-35 Com. Nal. Biodiversidad y CIQRO, México

Andréfouët, S. y Riegl, B. (2004). Remote sensing: a key tool for interdisciplinary assessment of coral reef precesses. Coral reef, Vol. 23, No. 1, (April 2004), pp 1-4, ISSN: 0722-4028

Andréfouët, S. ; Kramer, P. ; Torres-Pulliza, D. ; Joyce, K. ; Hochberg, E. ; Garza-Perez, R. ; Mumby, P. ; Riegl, B. ; Yamano, H. ; White, W. ; Zubia, M. ; Brock, J. ; Phinn, S. ; Naseer, A. ; Hatcher, B. & Muller-Karger, F. (2003). Multi-site evaluation of IKONOS data for classification of tropical coral reef environments. Remote sensing of environment, Vol. 88, No. 1-2. (November 2003), pp. 128-143, ISSN 0034-4257

Benfield, S., H. Guzman, J. Mair & A. Young (2007). Mapping the distribution of coral reefs and associated sublittoral habitats in Pacific Panama: a comparison of optical satellite sensors and classification methodologies. Internatinal joural of remote sensing, Vol. 28, No. 22, (November), pp. 5047-5070, ISSN 5047 5070

Brock, J., K. Yates & R. Halley (2006). Integration of coral reef ecosystem process studies and remote sensing, In: *Remote sensing of aquatic coastal ecosystem processes*, Richardson, L.L & LeDrew, E. F., 324 pp. Springer, ISBN-13 978-1-4020-3968-3, Netherlands

Call, A. K., T. J. Hardy & D. O. Wallin (2003). Coral reef habitat discrimination using multivariate spectral analysis and satellite remote sensing. *International Journal of remote sensing*, Vol. 24, No. 13, pp. 2627-2639

Camarena-Luhrs, T., (2003). Ficha Informativa de los Humedales de Ramsar: Reserva de la Biosfera Banco Chinchorro. Cancún, México.

Capolsini P, Andréfouët S, Cedric R, Payri C (2003) A comparison of Landsat ETM+, SPOT HRV, Ikonos, ASTER, and air- borne MASTER data for coral reef habitat mapping in South Pacific islands, *Canadian Journal of Remote Sensing*, Vol. 29, No. 2, pp. 187-200

Capolsini, P., B. Stoll, S. Andréfouët (2003). A comparison of classification algorithms for coral reef habitat mapping in South Pacific islands, IGARSS

Carricart-Ganivet J.P., E. Arias-González, G. García-Gil, G. Acosta-González, A. Beltrán-Torres, J. Castro-Pérez, N. Membrillo-Venegas y J. Padilla-Saldívar, (2002). Manejo integral de Banco Chinchorro: levantamiento geográfico y caracterización geomorfológica del arrecife, con especial énfasis en las comunidades coralina e íctica. SISIERRA, Mexico

Chávez, E. & Hidalgo, E. (1984). Spatial structure of benthic communities of Banco Chinchorro, México, In: *Advances in Reef Science, A Joint meeting of the Atlantic Reef Committee and the International Society for Reef Studies*, Miami

Clark, C. D., P.J. Mumby, J. R. M. Chisholm, J. Jaubert (2000). Spectral discrimination of coral mortality states following a severe bleaching. *International Journal of Remote Sensing*, Vol. 21, No. 11, (January 2000), pp. 2321-2327, ISSN 0143-1161

Elvidge, C.; Dietz J.; Berkelmans, R.; Andréfouët, S.; Skirving, William.; Strong, Alan. & Tuttle, B. (2004). Satellite observation of Keppel Islands (Great Barrier Reef) 2002 coral bleaching using IKONOS data. *Coral Reefs*, Vol. 23, No.1, pp 123-132, ISSN 07224028

García-Ureña, R. P., (2004). *Dinámica de crecimiento de tres especies de coral en relación a las propiedades ópticas del agua*, 24 September 2011, Available from: <http://grad.uprm.edu/tesis/garciaurena.pdf>

González, A.; Torruco, D.; Liceaga, A. & Ordaz, J. (2003). The shallow and deep bathymetry of the banco chinchorro reef in the mexican caribbean. *Bulletin of Marine Science*, Vol. 73, No. 1, pp. 15-22, ISSN 00074977

Gordon, H. (1978). Removal of atmospheric effects from satellite imagery of the oceans. *Applied Optics*, Vol. 17, No. 10, (May 1978) pp. 1631-1636

Green, E. P. ; Mumby, P. J. ; Edwards, A. J. & C. D., Clark, (1996). A review of remote sensing for the assessment and management of tropical coastal resources. *Coastal Management*, Vol. 24, No. 1, (January 1996), pp 1-40

Green, E., (2000). Satellite and airbone sensors useful in coastal applications. In: *Remote sensing handbook for tropical coastal management*, Green, E., P., Mumby, A. Edwards & C. Clark, Pp 41-56, UNESCO. ISBN 92-3-103736-6, France

Hedley, J.D. & Mumby, P.J. (2002). Biological and remote sensing perspectives on pigmentation in coral reefs. *Advances in Marine Biology*, Vol. 43, pp. 277-317,

Satellite Remote Sensing of Coral Reef Habitats Mapping in Shallow Waters at Banco Chinchorro Reefs,
México: A Classification Approach

37

Helge, J. ; Lindberg, B. ; Christensen, O. ; Lundälv, T. ; Svellingen, I. ; Mortensen, P. B. & Alvsvåg, J. (2005), Mapping of Lophelia reefs in Norway: experiences and survey methods, In: *Cold-water Corals and Ecosystems*, Freiwald, A. & Roberts J.M., pp 359-391, Springer-Verlag, ISBN 978-3-540-27673-9, Berlin Heidelberg

Hochberg, E. ; Atkinson, M. ; Apprill, A. & Andréfouët, S. (2004). Spectral reflectance of coral. *Coral Reefs*, Vol. 23. No. 1, (Abril 2004), Pp. 84-95, ISSN 0722-4028

Holden, H. (2002). Characterization of Optical Water Quality in Bunaken National Marine Park Indonesia. *Singapore Journal of Tropical Geography*, Vol. 23, No. 1, pp. 23-36, ISSN: 01297619

Holden, H. & LeDrew, E. (1998). Hyperespectral identification of coral reef features. *International Journal of Remote Sensing*, Vol. 20, No. 13, pp. 2545-2563

Hsu, M. K.; Liu, A. K.; Zhao, Y. & Hotta, K. (2008). Satellite remote sensing of Spratly Islands using SAR, *International Journal of Remote Sensing*, Vol. 29, No. 21, pp. 6427-6436, ISSN 01431161

Huston, M. (1985). Variation in coral growth rates with depth at Discovery Bay, Jamaica. *Coral Reefs*, Vol. 4, No. 1, (December 1985), pp. 19-25, ISSN 07224028

Jordán, E. (1979). Estructura y composición de los arrecifes coralinos en la región noreste de la península de Yucatán, México. *An. Centro de ciencias del mar y limnología, UNAM*, Vol. 16 No. 1, pp. 69-86

Karpouzli, E. ; Malthus, T. J. ; & Place, C. J. (2004). Hyperspectral discrimination of coral reef benthic communities in the western Caribbean, *Coral Reefs*, Vol. 23, (February 2004), pp. 141–151, ISBN 00338-003-0363-9

Kirk, J.T.O. (1994). *Light & photosynthesis in aquatic ecosystems* (2nd edition), Cambridge University Press, ISBN 0521 45966 4, New York

Krupa, S. (1999). *Polución, población y plantas*, Colegio de Postgraduados, México.

Kuhn, T., (2006). *La estructura de las revoluciones científicas*, Fondo de cultura económica, México

Lee, Z., K. L. Carder, C. D. Mobley, R. G. Steward & J. S. Patch (1999). Hyperspectral remote sensing for shallow waters: 2.Deriving bottom depths and water properties by optimization. *Applied Optics*, Vol. 38, No. 18, (June 1999), pp. 3831-3843

Lesser, M. P. & Mobley, C. D. (2007). Bathymetry, water optical properties, and benthic classification of coral reefs using hyperspectral remote sensing imagery. *Coral Reefs*, Vol. 26, No. 4, (January 2007) pp. 819-829, ISSN 07224028

Liceaga-Correa, M. A. & Euan-Avila, J. I. (2002). Assessment of coral reef bathymetric mapping using visible Landsat Thematic Mapper data. *International Journal of Remote Sensing*, Vol. 23, No. 1 pp 3-14, ISSN: 01431161

Loya, Y. (1972). Community structure and species diversity of hermatypic corals at Eilat, Red Sea. *Marine Biology*, Vol. 13, No. 2 pp. 100-123, ISSN 00253162

Lyzenga, D.R., (1981). Remote sensing of bottom reflectance and water attenuation parameters in shallow water using aircraft and Landsat data. *International Journal of Remote Sensing*, Vol. 2, No. 1, (Enero 1981), pp. 71-82. ISSN 0143-1161

Maritorena, S., A. Morel & B. Gentili (1994). Diffuse Reflectance of Oceanic Shallow Waters: Influence of Water Depth and Bottom Albedo. *Limnology and Oceanography*, Vol. 39, No. 7 (Nov., 1994), pp. 1689-1703

Mobley, C.D., B. Gentili, H. R. Gordon, Z. Jin, G. W. Kattawar, A. Morel, P. Reinersman, K. Stamnes & R. H. Stavn (1993). Comparison of numerical models for computing underwater light fields. *Applied Optics*, Vol. 32, No. 36, (December 1993) pp. 7484-7504

Mumby, P.J, E.P. Green, A.J. Edwards & C.D. Clark (1997). Coral reef habitat mapping: how much detail can remote sensing provide? *Marine Biology*, Vol. 130, No.4 (December 1997), pp. 193-202. ISSN: 0025-3162

Mumby, P.J. & Harborne, A.R. (1999). Development of a systematic classification scheme of marine habitats to facilitate regional management of Caribbean coral reefs. *Biological Conservation*, Vol. 88, No. 2, (May 1999), pp155-163, ISSN 0006-3207

Mumby, P.J., W. Skirving, A.E. Strong, J.T. Hardy, E.F. LeDrew, E.J. Hochberg, R.P. Stumpf & L.T. David. (2004). Remote sensing of coral reefs and their physical environment. *Marine Pollution Bulletin*, Vol.48, No. 3-4, (February 2004), pp. 219-228, ISSN 0025-326

Mumby, P.J., A.R. Harborne, J.D.Hedley, K. Zychaluk & P. Blackwell (2006). The spatial ecology of Caribbean coral reefs: revisiting the catastrophic die-off of the urchin Diadema antillarum. *Ecological Modelling*, Vol. 196, pp. 131-148

Newman, C., E. LeDrew & A. Lim. (2006). Mapping of coral reef for management of marine protected areas in developing nations using remote sensing. In: *Remote sensing of aquatic coastal ecosystem processes*, Richardson, L. y LeDrew, E., pp. 251-278, Springer, ISBN 978-1-4020-3968-3, Netherlands

Pahlevan, N.; Valadanzouj, M. J. & Alimohamadi, A. (2006). A quantitative comparison to water column correction techniques for benthic mapping using high spatial resolution data. *ISPRS Commission VII Mid-term Symposium "Remote Sensing: From Pixels to processes"*, Enschede, Netherlands, 8-11 May 2006

Philpot, W. D. (1989). Bathymetric mapping with passive multispectral imagery. *Applied Optics*, Vol. 28, No. 8 (April 1989), pp. 1569-1578

Primack, R., R. Rozzi, P. Feinsiger, R. Dirzo y F. Massardo. (1998). *Fundamentos de conservación biológica*. Fondo de cultura económica. México.

Richter, R. (1996). Atmospheric correction of satellite data with haze removal including a haze/clear transition region. *Computers & Geosciences*, Vol. 22, No. 6, pp. 675-681, ISSN 00983004

Richter, R. (1998). Correction of satellite imagery over mountainous terrain. *Applied Optics*, Vol. 37, No. 16, (June 1998), pp. 4004-4015

Rivera, J. A. ; Prada, M. C. ; Arsenault, J.L. ; Moody, G. & Benoit, N. (2006). Detecting fish aggregations from reef habitats mapped with high resolution side scan sonar imager, In: *Emerging technologies for reef fisheries research and management*, NOAA, pp. 88-104. NOAA Professional Paper NMFS, 5 Seattle, WA.

Secretaría de Marina, Dirección General adjunta de Oceanografía, Hidrografía y Meteorología (SEMAR) (2008). *Carta náutica SM-932 Majahual to Banco Chinchorro*, Scale 1: 100000

SERC (September 2011). Phytoplankton lab, 10.09.2011 Available from http://www.serc.si.edu/labs/phytoplankton/primer/hydrops.aspx

Xin, Y., J. Li y Q. Cheng (2007). Automatic Generation of Remote Sensing Image Mosaics for Mapping Large Natural Hazards Areas, In: *Geomatics solutions for disaster management*, Li, J., Zlatanova, S. y Fabbri, A. 451 pp. Springer, ISBN 13 978-3-540-72106-2, Springer Berlin Heidelberg New York.

# 3

# Remote Sensing of Cryosphere

Shrinidhi Ambinakudige and Kabindra Joshi
*Mississippi State University*
*USA*

## 1. Introduction

The cryosphere is the frozen water part of the Earth's system. The word is derived from the Greek "kryos," meaning cold. Snow and ice are the main ingredients of the cryosphere and may be found in many forms, including snow cover, sea ice, freshwater ice, permafrost, and continental ice masses such as glaciers and ice sheets. Snow is precipitation made up of ice particles formed mainly by sublimation (NSIDC, 2011). Ice is the key element in glaciers, ice sheets, ice shelves and frozen ground. Sea ice forms when the ocean water temperature falls below freezing. Permafrost occurs when the ground is frozen for a long period of time, at least two years below $0^0$ C, and varies in thickness from several meters to thousands of meters (NSIDC, 2011). Glaciers are thick masses of ice on land that are caused by many seasons of snowfall. Glaciers move under their own weight, the external effect of gravity, and physical and chemical changes. The cryosphere lowers the earth's surface temperature by reflecting a large amount of sunlight, stores fresh water for millions of people, and provides habitat for many plants and animals.

Apart from the Arctic and Antarctic regions, the cryosphere is mainly a high altitude phenomenon. It is found on Mount Kilimanjaro in Africa, the Himalayan mountain range, high mountains of United States, and in Canada, Russia, Japan, and China. Researchers in the cryosphere are often hindered by the lack of accessibility due to the rugged terrain. In such cases, remote sensing technologies play an important role in cryosphere research. These techniques are imperative for researchers studying glacial retreat and mass balance change in relation to global climate change.

The cryosphere has a significant influence on global climate and human livelihoods. Change in spatial and temporal distribution of the cryosphere influences the water flow in the world's major rivers. Among the various parts of the cryosphere, glaciers play the most important role in climate change studies since glacier recessions are indicators of global climate change (Oerlemans et al., 1998; Wessels et al., 2002; Ambinakudige, 2010). Retreating glaciers can pose significant hazards to people (Kaab et al., 2002). Glacier retreat often lead to the formation of glacial lakes at high altitudes, the expansion of existing lakes, and the potential for glacial lake outburst floods (GLOFs) (Fujita et al., 2001; Bajracharya et al., 2007). A GLOF is the sudden discharge of a huge volume of water stored in a glacial lake due to huge ice falls, earthquakes, avalanches, rock fall or failure of a moraine dam (Grabs & Hanisch, 1993). There are more than 15,000 glaciers and 9,000 glacial lakes in the Himalayan mountain ranges of Bhutan, Nepal, Pakistan, China and India (Bajracharya et al., 2007). All these countries within the Himalayan region have at some time or another suffered a flood from a glacial lake outburst causing loss of property and lives, and these floods can be disastrous for the downstream riparian area (Richardson & Reynolds, 2000; Bajracharya et al., 2007). The significance of the glaciers as fresh water resources for millions of people is another reason to justify the continuous monitoring of these glaciers (Shiyin et al., 2003). Therefore, monitoring glaciers has significant importance both in understanding global climate change and in sustaining the livelihoods of the people downstream of the glaciers.

This chapter explores the use of remote sensing technologies in studies of the cryosphere and particularly in glaciers. First, we will discuss remote sensing sensors that are effective in monitoring glaciers. Then we will discuss the global effort to create glacier data, using remote sensing tools to delineate glacier areas, estimate volume and mass balance.

## 2. Remote sensors for monitoring glaciers

Remote sensing methods are more convenient than field methods to measure changes in glaciers. Studies have used Landsat (Ambinakudige, 2010), SPOT (Berthier et al., 2007), Terra ASTER (Rivera & Casassa, 1999; Kaab, 2007), IRS (Kulkarni et al. 2011), and ALOS (Narama et al., 2007) sensors successfully to measure glacial parameters. High-resolution satellite data such as IKONOS (Huggel et al., 2004) and Quickbird (Schmidt, 2009) have also been used in studies on glaciers. ASTER, SPOT5, IRS-1C, Resourcesat1 and 2, CORONA KH-4, KH 4A and KH 4B satellites have the capability to acquire stereoscopic images from which elevation data can be extracted for monitoring glacial surfaces in three dimensions (Racoviteanu, et al. 2008). Digital elevation models developed from these stereoscopic images can be used in measuring volume and mass balance change in glaciers.

### 2.1 Glacial mapping

Glacial mapping using remote sensing initially involved manual digitization of glacier boundaries on a false color composite (FCC) of Landsat MSS and TM images in Iceland (Williams, 1987) and Austria (Hall et al., 1992). Figure 1, is a picture taken by the astronauts in the international space station shows the snow and ice in Colonia glacier and its surroundings in Chile (International Space Station, 2000).

Snow has high reflectivity in the visible wavelength region and relatively less reflectivity in middle and shortwave infrared regions (Pellika and Rees, 2010). Freshly fallen snow has the highest reflectance in the visible and near-infrared wavelengths. Firn (partially compacted snow) has 25-30 % less reflectance than snow. Ice in glaciers has high reflectance in blue (400- 500 nm) and green (500 – 600 nm) wavelengths. However, in red (600-700 nm) the reflectance of ice is near zero. Debris in glaciers will significantly reduce the reflectance (Pellika and Rees, 2010). The spectral reflectance properties of snow also depend on time and season of the year. Sharp changes in the reflectance of snow can be seen when the melted snow recrystallizes to form firn and the density of snow changes. Similarly, albedo, which is the ratio of radiation reflected from a surface to the radiation incident on that surface, also varies among different cryospheric surfaces. Snow has high albedo (0.8 – 0.97) while dirty ice has low albedo (0.15-0.25). A high albedo value helps to reflect a huge amount of sunlight, which otherwise would have heated the earth's surface. Snow and ice reflectance is the main characteristic that is measured using remote sensing techniques. This

ISS01E5107 2067/25/45 00:00:00

Fig. 1. Astronaut photograph of Colonia Glacier, Chile. Photographer: International Space Station (2000).

characteristic assists in delineating glacier boundaries and classifying various cryospheric surface types (Pellika and Rees, 2010).

In figure 2, the Landsat TM bands 1 to 6, acquired on 25 April 2010, are shown to compare spectral characteristics of glaciers. This figure presents the area around the Imja glacier in the Sagarmatha National Park in the Himalayas of Nepal. TM1 (0.45 – 0.52 μm) is useful to distinguish snow/ice in cast shadow, and also in mapping glacier lakes. Snow and firn areas get saturated in TM1. TM2 (0.52 – 0.60 μm) and TM3 (0.63 – 0.69 μm) have very similar spectral reflectance. TM2 is also useful in distinguishing snow and ice in cast shadow. TM3 is used in band ratio such as the normalized difference vegetation index (NDVI), which helps in classification of ice and snow in areas of dense vegetation. TM4 (0.76-0.90 μm) in the near infrared wavelength region has less reflectance from snow than TM2 and TM3. The clean ice region looks darker in near infrared band, indicating lower reflectance due to the presence of water at the surface (Hall et al., 1988). In TM5, the snow-covered area absorbs nearly all radiation and appears almost dark. This band is also useful in identifying clouds. The thermal band TM6 (2.08-2.35 μm) registers thermal emissions from the surface. Debris has a higher temperature and thus brighter pixels. Thick debris on ice can be easily distinguished as it will have a higher temperature (Pellika and Rees, 2010).

The high reflectance of the snow compared to the ice makes it easy to separate snow and ice. Snow and clouds are often difficult to distinguish when single imagery is used. Clouds and snow have similar reflectance at wavelengths below 1 μm in the near infrared region. The distinction between snow and ice is clearer near 1.55 and 1.75 μm. Therefore a ratio of two

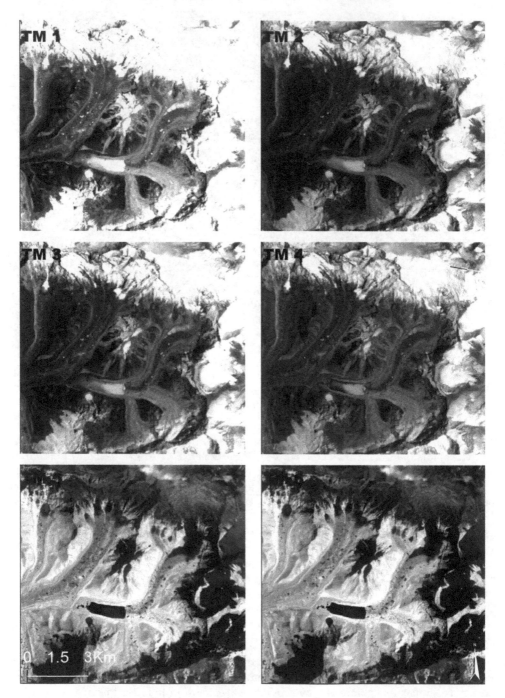

Fig. 2. Six Landsat TM bands showing Imja and surrounding glaciers in the Himalayas.

spectral bands in the visible (VIS) and shortwave near infrared (SWIR) regions are commonly used in automated mapping of snow and ice. A Normalized Difference Snow Index (NDSI) calculated as (VIS-SWIR)/(VIS+ SWIR) helps to separate snow and ice from darker areas such as rocks and soils. Whereas the visual spectrum band covers a wavelength of 0.57µm, the shortwave infrared band covers 1.65µm. If the NDSI value exceeds 0.4, it is assumed there is snow cover on the ground (Dozier 1984). However, seasonal variation in NDSI value for snow has also been observed. A threshold of 0.48 in July and 0.6 in September was observed as an optimal threshold during the field investigation in Abisko, Sweden (Vogel 2002). The Normalized Difference Water Index (NDWI) calculated as (NIR-VIS)/(NIR+VIS) is useful to differentiate water from snow, ice and other physical features. NDWI is very useful in detecting formation of new, as well expansion of existing, glacial lakes (Huggel et al., 2002).

Band ratios such as Landsat TM3/TM5 or TM4/TM5 are also helpful in mapping glacial areas. The TM4/TM5 ratio is more appropriate for clean-ice glacier mapping (Paul and Kaab, 2001), whereas TM3/TM5 works better in areas of dark shadow and thin debris cover (Andreassen et al., 2002). Both NDSI and ratio methods have similar robust outcomes in glacier mapping and are recommended. NDSI and band ratio methods often misclassify debris-covered glacier ice because of the similarity in spectral signature to the surrounding debris Band ratio VIS/NIR also often misclassifies proglacial lakes. The band ratio NIR/SWIR is good for classifying only clean glacier ice (Bhambri and Bolch, 2009). Therefore, manual corrections have to be made after classifying glaciers using either NDSI or ratio methods. High resolution panchromatic images used in manual corrections can help to delineate the precise boundaries of glaciers. Similarly, figure 3 also indicates spectral responses of different types of snow and water.

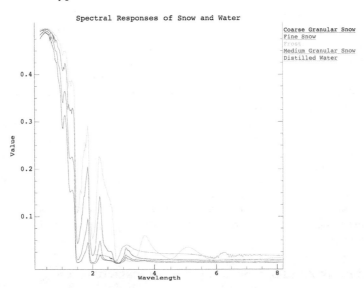

(Data Source: ENVI Spectral Library derived from John Hopkins University
http://asterweb.jpl.nasa.gov/speclib/)

Fig. 3. Spectral responses of different types of snow and water

## 3. Digital elevation models in glacier mapping

Debris-covered glaciers can be mapped effectively by using morphometric parameters derived from DEM and thermal bands (Ranzi et al., 2004). The surface temperatures of the debris on glacier ice are cooler than the debris outside the ice, which can be identified by a thermal image. AVHRR and MODIS satellites have coarse resolution thermal bands; the Landsat TM/ETM+ satellite, on the other hand, has a higher resolution thermal band. Using morphometric characteristics like slope can help to separate a debris-covered glacier from its surrounding moraines (Bishop et al., 2001). DEMs help in estimating morphometric characteristics such as slope, glacier profile curvature, and aspect. DEMs are most important for the estimation of volume change for inaccessible glacier regions (Bolch et al., 2008).

DEMs are generated from topographic maps, spaceborne optical stereo data, synthetic aperture radar (SAR) imagery, spaceborne radar and laser altimetry, such as LiDAR data. Terra ASTER optical stereo satellite data with a spatial resolution of 15 m has been used successfully in generating DEM to study glaciers (Kargel et al., 2005). SPOT-5, ALOS PRISM, CartoSat I and II, IKONOS, Quickbird and GeoEye-1 satellites also have stereo capability and can be used in creating DEMs. ASTER and ALOS PRISM produce along-track stereo images and are capable of simultaneous image acquisition. Other sensors provide across-track acquisition with a time lag, which causes problems in certain situations, such as under different atmospheric conditions (Bhambri & Boch, 2009). Measuring surface changes using DEMs, known as indirect geodetic methods, is a useful method to calculate a glacier's volume and mass balance changes (Etzelmüller & Sollid, 1997).

Another technology used to study the cryosphere is radar, which is an active microwave system composed of a transmitter and a receiver. Radar imagery is not confined only to daylight, cloud-free conditions, and thus has advantages in estimating glacial change. The first spaceborne SAR was the SEASAT satellite, launched in 1978. ERS-1/2, Envisat – ASAR, Radarsat-1, ALOS PALSAR, Radarsat-2 and TerraSAR-X are some of the satellites with spaceborne SAR launched after the initial success of SEASAT. Due to its all-day, all-weather imaging capabilities, large spatial coverage, and ability to measure minute changes on the earth's surface, SAR has a major advantage in glacier studies. Digital elevation models (DEM) created using SAR images are useful in measuring glacier mass balance, glacier velocity, and snow density.

Light Detection and Ranging (LiDAR) is also used in remote sensing of glaciers. High density measurement with a high vertical accuracy (10 cm) allows a very detailed representation of terrain in LiDAR imagery. LiDAR imagery also comes with accurate positional information because of the global positioning system receiver in the LiDAR instruments on aircraft.

## 4. The geodetic method of measuring glacier mass balance

The mass balance of a glacier is referred to as the total loss or gain in glacier mass at the end of a hydrological year (Cuffey and Paterson, 2010). Glacier mass balance is the link between climate and glacier dynamics (Kaser et al., 2002). Mass balance change is a direct reaction to atmospheric conditions. The specific mass balance can be compared directly between different glaciers. This makes it easier than using length changes to establish a link to climate data (UNEP, 2008). Glacier mass losses affect local hydrology and are thus important for regional water supplies and assessing global sea level rise.

In the geodetic method, changes of glacier volume are measured from maps and elevation models, unlike the direct or glaciological methods that derive mass changes from ground-based spot measurements. With the development of remote sensing techniques, digital elevation models (DEM) are commonly used in the geodetic method. The difference in DEM values in two years is multiplied by the glacier accumulation area to obtain volume change (dV). The difference in volume (dV) multiplied by the density of ice, generally considered to be 850-900 Kg/m$^3$, would provide the mass balance (Cuffey & Paterson, 2010).

The accumulation and ablation area in a typical glacier is separated by an equilibrium line altitude (ELA). This line divides the accumulation zone (the higher reaches of a glacier where there is a net mass gain) and the ablation zone (the lower reaches where mass is lost). ELA is the elevation at which mass is neither gained nor lost through the course of a hydrological year. At the end of the season, the elevation at which there was no net gain or loss is identified as the ELA. The snow line altitude (SLA) divides the ice in the ablation zone from the snow in the accumulation zone. Since ice and firn have lower albedo, SLA can be determined using remote sensing images (Khalsa et al., 2004; Racoviteanu et al., 2007).

The geodetic approach has been used in several studies based on historical topographic maps and DEMs derived from SPOT imagery (Berthier et al., 2007), SRTM (Racoviteanu et al., 2007), ASTER (Rivera & Cassassa, 1999; Kaab, 2007). Studies have also used high resolution DEMs derived from ALOS PRISM and Corona (Narama et al., 2007) to estimate mass balances with the geodetic method.

The two bands in ASTER VNIR, 3N and 3B, generate an along-track stereo pair with a base-to-height (B/H) ratio of about 0.6. Studies have found that the DEM accuracy has a linear relation with terrain slopes (Toutin, 2008; Bolch et al., 2004; Racoviteanu et al., 2007). DEMs created using ASTER images on Mt. Fuji, Japan; in the Andes Mountains, Chile-Bolivia; at San Bernardino, CA and Huntsville, AL resulted an accuracy of ±5 m, ±10m, ± 6m and ± 1.5m respectively (Hirano, 2003). Therefore, accuracy, number and distribution of GCPs are required to create an accurate DEM.

The ALOS data have been available since January 2006. ALOS has three remote-sensing instruments. The Panchromatic Remote-Sensing Instrument for Stereo Mapping (PRISM) sensor of ALOS data is suitable for digital elevation mapping. It consists of three independent telescopes for forward, nadir and backward view, and each telescope provides 2.5m spatial resolution. The Advanced Visible and Near Infrared Radiometer type 2 (AVNIR-2) sensor collects data with 10m resolution and is suitable for glacier mapping. The accuracy of elevation extracted from ALOS PRISM is 5 m (Racoviteanu et al., 2008).

The three images (forward, nadir and backward) that ALOS provides are required for creation of DEM (Ye, 2010). The base-to-height (B/H) of PRISM on ALOS is set to 1.0

(forward view + backward view) and 0.5 (sidelong view + nadir view). ALOS images have been used to produce highly accurate DEMS. The DEMs created for the northern slope of Qomolangma in the Mt. Everest region had a mean elevation difference of 1.7m with a DEM created using topographic maps in non-glaciated areas. The mean difference between Aster and ALOS images was found to be about 45m (Ye, 2010).

Once the outlines of the glaciers are delineated, they can be combined with DEM to derive glacier parameters such as length, termini elevations, and volume. DEMs derived from SPOT5, ASTER, CORONA or ALOS PRISM can be used in mass balance studies.

## 5. Global Land Ice Measurements from Space (GLIMS)

In an effort to analyze the glacial change throughout the world, a global-level consortium, the Global Land Ice Measurements (GLIMS), has established a database at the National Snow and Ice Data Center (NSIDC) in Boulder, Colorado (Raup et al., 2007). Under GLIMS, 12 regional centers are working to acquire satellite images, analyze them for glacial extent and changes, and assess change data for causes and implications for people and the environment.

GLIMS is an international consortium established to monitor the world's glaciers. Although GLIMS is making use of multiple remote-sensing systems, ASTER (Advanced Spaceborne Thermal Emission and Reflection Radiometer) satellite images are the major data input in GLIMS database. The GLIMS team has put together a network of international collaborators who analyze imagery of glaciers in their regions of expertise. Collaborators provide digital glacier outlines and metadata. Data also include snow lines, center flow lines, hypsometry data, surface velocity fields, and literature references. The National Snow and Ice Data Center archives the data provided by the regional centers.

The GLIMS team also developed tools to aid in glacier mapping, such as GLIMSView, which is an open-source, cross-platform application designed to support and standardize the glacier digitization process. It allows regional centers to transfer data to the National Snow and Ice Data Center for incorporation into the GLIMS glacier database. Users can view various types of satellite imagery, digitize glacier outlines and other material units within the images, attach GLIMS-specific attributes to segments of these outlines, and save the outlines to ESRI shapefiles. GLIMSView is free and available at http://www.glims.org/glimsview/.

## 6. Conclusions

Satellite remote sensing of the cryosphere has progressed over the last five decades. It has helped us to understand the global distribution of the cryosphere, variation and trends in snow cover, sea ice, and glaciers. We have a pretty decent map of the cryosphere. Remote sensing has helped in rapid assessment of glaciers in hostile ground conditions in areas such as Antarctica, the Artic and alpine glaciers.

There are several challenges in remote sensing of the cryosphere. Acquiring cloud-free satellite imagery is still challenging. Synthetic aperture radar (SAR) imagery has received a great deal of attention in recent years as it can provide cloud-free data. SAR interferometry has been used successfully in areas such as glacier motion and topographical mapping. The use of radar has been gaining more attention recently. Ground penetrating radars are being used to study the internal structure and bedrock configuration of glaciers.

For most part, *in situ* measurement of the cryosphere is often not a viable option, so the focus of cryosphere study remains on the use of remote sensing techniques. The World Glacier Monitoring Service (WGMS) coordinates the global glacier observation strategy with the help of the Global Land Ice Measurement from Space project and the European Space Agency's Global Glacier Project.

Spaceborne remote sensing techniques in the last five decades have shown tremendous advancement. From Landsat to InSAR imagery, the remote sensing technology has helped in understanding and mapping the cryosphere. Many of these data are available free or for low cost; some of them are very expensive, and using them requires specialized skill. With the increase in computer processing power, the potential for the collection, storage, transmission and processing of remotely sensed data on the cryosphere has improved.

Changes in glaciers provide evidence of climate change, and therefore glaciers play a key role in early detection of global climate-related observations (WGMS, 2011). Glacier change will impact global sea level fluctuations and other natural hazards. These environmental changes require international glacier monitoring efforts to make use of remote sensing and geo-informatics along with the more traditional field observations.

## 7. References

Ambinakudige, S. (2010). A study of the Gangotri glacier retreat in the Himalayas using Landsat satellite images. *International Journal of Geoinformatics* 6 (3), pp. 7-12.

Andreassen, L.M., Elvehoy, H. & Kjollmoen, B. (2002). Using aerial photography to study glacier changes in Norway. *Annals of Glaciology* 34: pp. 343-348.

Bajracharya, S.R., Mool, P.K., & Shrestha, B. (2007). *Impact of climate change on Himalayan glaciers and glacial lakes: case studies on GLOF and associated hazards in Nepal and Bhutan.* Kathmandu, International Centre for Integrated Mountain Development and United Nations Environmental Programme Regional Office Asia and the Pacific. (ICIMOD Publication 169).

Berthier, E., Arnaud, Y., Kumar, R., Ahmad, S., Wangnon, P., & Chevallier, P. (2007). Remote sensing estimates of glacier mass balances in the Himachal Pradesh (Western Himalaya, India). *Remote Sensing of Environment* 108:327-338.

Bhambri, R., & Bolch, T. (2009). Glacier Mapping: A Review with special reference to the Indian Himalayas, *Progress in Physical Geography* 33(5), 672–704.

Bolch, T., Buchroithner, M. F., Pieczonka, T., & Kunert, A. (2008). Planimetric and volumetric glacier changes in Khumbu Himalaya since 1962 using Corona, Landsat TM and ASTER data, *Journal of Glaciology*, 54, pp. 592–600.

Bishop, M.P., Bonk, R., Kamp, U. & Shroder, J.F. (2001): Topographic analysis and modeling for alpine glacier mapping. *Polar Geography*, 25: 182-201.

Cuffey, K.M & Paterson, W.S.B. (2010). *The Physics of Glaciers.* 4th ed. Academic Press.

Dozier, J., (1984). Snow reflectance from Landsat-4 Thematic Mapper. *IEEE Transactions on Geoscience and Remote Sensing.* 22(3), pp. 323-328.

Etzelmüller, B. & Sollid, J.L. (1997). Glacier geomorphometry - an approach for analysing long-term glacier surface changes using grid-based digital elevation models. *Annals of Glaciology*, 24, pp. 135-141.

Fujita, K. Kadota, T. Rana, B., Kayastha, R.B., & Ageta, Y. (2001). Shrinkage of glacier Ax010 in Shorong region, Nepal Himalayas in the 1990s. *Bulletin of Glacier Research* 17, pp. 51-54.

Grabs, W. & Hanisch, J. (1993). Objectives and prevention methods for glacier lake outburst floods (GLOFs). *Snow and Glacier Hydrology*: Proceedings of the Kathmandu symposium, November 1992. IAHS Publication, 218.

Hall, D.K., Williams, R.S., & Bayr, K.J. (1992). Glacier récession in Iceland and Austria. *EOS*, 73(12), 129.

Hirano, A. (2003). Mapping from ASTER stereo image data: DEM validation and *accuracy* assessment. *ISPRS Journal of Photogrammetry and Remote Sensing*, 57(5-6), pp. 356-370.

Huggel, C., Kääb, A., Haeberli, W., Teysseire, P. & Paul, F. (2002). Remote sensing based *assessment* of hazards from glacier lake outbursts: a case study in the Swiss Alps. *Canadian Geotechnical Journal*, 39, pp.316-330.

Huggel, C., Kääb, A., & Salzmann, N. (2004). GIS-based modeling of glacial hazards and their interactions using Landsat-TM and IKONOS imagery. *Norsk Geografisk Tidsskrift - Norwegian Journal of Geography*, 58(2), pp 61-73.

International Space Station. (2000). Colonia glacier: from the glacier photograph *collection*. Boulder, Colorado USA: National Snow and Ice Data Center/World Data Center for Glaciology. Digital media. http://eol.jsc.nasa.gov/scripts/sseop/photo.pl?mission=ISS001&roll=E&frame=5107

Kargel, J.S, Abrams, M., Bishop, M., Bush, A., Hamilton, G., Jiskoot, H., Kaab, A., Kieffer, H.H., Lee, E., Frank, P., Rau, F., Raup, B., Shroder, J.F., Soltesz, D., Stainforth, D., Leigh, S, & Wessels, R. (2005). Multispectral imaging contributions to global land ice measurements from space. *Remote Sensing of Environment*, 99(1-2), pp.187-219.

Kaab, A. (2007). Glacier volume changes using ASTER optical stereo. A test study in Eastern Svalbard. *IEEE Transactions on Geosciences and Remote Sensing* 10:3994 -3996.

Kaab, A, Paul, M. Maisch, M., Hoelzle, M. & Haeberli, W. (2002). The new remote sensing derived Swiss glacier inventory: II. first results. *Annals of Glaciology*, 34.

Kaser, G., Fountain, A., & Jansson, P. (2002). A manual for monitoring the mass balance of mountain glaciers. *International Hydrological Programme*, 69.

Khalsa, S.J.S., Dyurgerov, M.B., Khromova, T., Raup, B. H., & Barry, R.G. (2004). Space-based mapping of glacier changes using aster and gis tools. *IEEE transactions on geosciences and remote sensing*, 42, pp. 2177-2183.

Kulkarni, A.V., Rathore, B.P., Singh, S. K., & Bahuguna, I. M. (2011). Understanding changes in the Himalayan cryosphere using remote sensing techniques. *International journal of remote sensing*, 32: 3, 601-615.

Narama, C., Kaab, A., Kajiura, T., & Abdrkhmatov, K. (2007). Spatial variability of recent glacier area and volume changes in central Asia using corona, Landsat, Aster and ALOS optical satellite data. *Geophysics Research Abstract*, 9.

NSIDC (2011). Education resources. http://nsidc.org/pubs/education_resources/

Oerlemans, J., Anderson, B., Hubbard, A., Huybrechts, P., Jóhannesson, T., Knap, W. H., Schmeits, M., Stroeven, A. P., van de Wal S.R.W., & Wallinga, J. (1998). Modelling the response of glaciers to climate warming. *Climate Dynamics* 14, pp.267-274.

Paul, F & Kaab, A. (2005). Perspectives on the production of a glacier inventory from multispectral satellite data in the Canadian Arctic : Cumberland Peninsula, Baffin Island. *Annals of Glaciology*. 42, pp. 59-66.

Pellika P. & Rees, W.G., (2010). *Remote Sensing of Glaciers*. CRC Press. London.

Ranzi, R., Grossi, R., Iacovelli, l. & Taschner, T. (2004). Use of multispectral ASTER images for mapping debris-covered glaciers within the GLIMS Project. *IEEE International Geoscience and Remote Sensing Symposium* 2, 1144-1147.

Racoviteanu, A., Manley, W.F., Arnaud, Y. & Williams, M. (2007). Evaluating digital elevation models for glaciologic applications: An example from Nevado Coropuna, Peruvian Andes. *Global and Planetary Change*, 59(1-4), pp.110-125.

Racoviteanu, A.E., Williams, M.W., & Barry, R.G. (2008). Optical Remote Sensing of Glacier Characteristics: A Review with Focus on the Himalaya. *Sensors*, 8, pp 3355-3383.

Raup, B., Racoviteanu, A., Khalsa, S. J. S., Helm, C., Armstrong, R. & Arnaud, Y. (2007). The GLIMS geospatial glacier database: A new tool for studying glacier change. *Global and Planetary Change*, 56, pp. 101-110.

Richardson, S.D. & Reynolds. J.M. (2000). An overview of glacial hazards in the Himalayas. *Quaternary International*, 65-66, pp.31-47.

Rivera, A. & Casassa, G. (1999). Volume changes on Pio Xi Glacier, Patagonia. *Global Planet Change* 22:233-244.

Schmidt, S. (2009). Fluctuations of Raikot Glacier during the past 70 years: a case study from the Nanga Parbat massif, northern Pakistan. *Journal of Glaciology*, 55 (194), pp. 949-959.

Shiyin, L., Wenxin, S., Yongping, S., & Gang, L. (2003). Glacier changes since the little ice age maximum in the western Qilian Shan, Northwest China, and consequences of glacier runoff for water supply. *Journal of Glaciology*, 49, pp.117-124.

Toutin, T. (2008). ASTER DEMs for geomatic and geoscientific applications: a review. *International Journal of Remote Sensing*, 29(7), pp. 1855-1875.

UNEP. (2008). *World Heritage Site: Sagarmatha National Park, Nepal*. United Nations Environment Programme.

Vogel, S.W. (2002). Usage of high-resolution Landsat-7 band 8 for single band snow cover classification. *Annals of Glaciology*, 34, pp.53-57.

WGMS (2011). World glacier monitoring service. http://www.wgms.ch/

Wessels, R., Kargel, J. S. & Kiefffer. H.H. (2002). Aster measurement of supraglacial lakes in the Mount Everest region of the Himalaya. *Annals of Glaciology*, 34, pp. 399-408.

Williams, R. (1987). Satellite remote sensing of Vatnajökull, Iceland, *Annals of Glaciology*, 9, pp. 119-125.

Ye, Q. (2010). The generation of DEM from ALOS/PRISM and ice volume change in Mt. Qomolangma region. *Geophysical Research Abstracts*, 12.

# 4

# Remote Sensing and Environmental Sensitivity for Oil Spill in the Amazon, Brazil

Milena Andrade[1] and Claudio Szlafsztein[2]
[1]Federal University of Pará, Amazon Advance Studies (NAEA)
[2]Federal University of Pará, Center of Environment (NUMA)
Brazil

## 1. Introduction

The use of remote sensing has become a fundamental tool for the identification and analysis of different types of risks in coastal zones. The numerous and, in some cases, recent incidents of oil spills have encouraged companies and government agencies to improve methods, both anticipatory and corrective, to minimize damages. The term 'risk' denotes the possibility that adverse effects may occur as a result of natural events or human activities (Kates et al., 1985). Risk is defined as an association between the hazard´s characteristics (e.g. frequency, magnitude and location) and the vulnerability of affected human populations, environment and infrastructure (Wisner et al., 2004). Risk can be classified by their origin, such as natural, social, or technological (Renn, 2008). Oil spills are an example of the last category, and the coastal areas are one of the most impacted. Environmental sensitivity to oil impacts can be defined through the coastal Environmental Sensitivity Index (ESI), which considers: (i) the geomorphologic aspects such as type and slope of coastline and the degree of exposure to the energy of waves and tides; (ii) oil sensitive biological resources; and (iii) the socio-economic activities that can be affected by oil spills (Gundlach & Hayes, 1978; Dutrieux et al., 2000).

In Brazil, environmental sensitivity mapping has been carried out under the law 9966/2000, which gave the Ministry of the Environment (Climate Change and Environmental Quality Secretary) responsibility to identify, locate and define the boundaries of ecologically sensitive areas with respect to the spill of oil and other dangerous substances in waters within national jurisdiction. This way, based on PETROBRAS (2002) and NOAA (2002), the specifications and technical standards for preparing environmental sensitivity maps for oil spills in coastal and marine zones was elaborated upon (MMA, 2002). Such environmental sensitivity maps provide information in an easy format being useful to determine priorities to impact protection and mitigation. Identification and mapping is developed at three levels: (i) Strategic (1:500,000 for the entire area of a hydrographical basin); (ii) Tactical (1:150,000 for the entire coastline mapped); and (iii) Operational (up to 1:50,000 for a high-risk/sensitivity areas). Each of these mapping scales uses specific tools for remote sensing and GIS tools.

The Amazonian coastal zone extends along ~2250 km, not including the several inlets, islands and small estuaries, which punctuate the coastline (Souza Filho et al., 2005a). This coastal zone is placed in the context of the tropical humid regions, in a low-lying area with active processes of erosion, sedimentation and neotectonics. Also, it is marked by a great hydrologic influence; in a meso- to macrotidal area (Souza Filho, 2005). It is a high-density drainage network, in which the Amazon River discharges a volume of water of 6.3 trillion m³/year and of sediment estimated at 1.2 billion tons/year (Meade et al., 1985).

Such environmental characteristics are responsible for the development of an extensive mud plain and mangrove area which is located in three States (Amapá, Pará and Maranhão), is approximately 8,386 km² wide, and contains 80% of all mangroves in Brazil (Herz, 1991). Where macrotides are present, the area of a flooded mangrove may extend for up to 30 km inland, and the estuaries themselves as much as 80 km (Souza Filho, 2005) (Figure 1). These extensive mud and mangroves plains are considered to be one of the most sensitive areas to oil spills. Also, these mangroves are along national and international ships routes. Transportation and storage are mainly responsible for oil spills in Amazonian coastal zone, since there is no expressive exploration. In 2001, in the state of Pará, approximately 1900 tons of oil sank near the Port of Vila do Conde (Berredo et al., 2001).

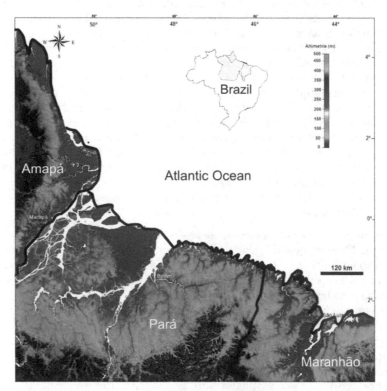

Fig. 1. Amazonian coastal zone in radar SRTM representation (source: modified from Souza Filho et al., 2005a)

In this sense, researches from Federal University of Pará have been working on several projects since 2001 aiming to study the Amazonian coastline and the impact of oil spills on the environment. Therefore, from 2004 to 2010, a large group of scientists were grouped in PIATAM-Mar project "Potential Environmental Impacts and Risks of the Oil and Gas Industry", financially supported by PETROBRAS[1], to map and analyse the vulnerability of the Amazonian coastline oil related disasters. Since 2012, the project "Elaboration of Environmental Sensitivity Maps (SAO maps) for oil spills in Pará-Maranhão and Barreirinhas Basins", founded by the National Research Council of Brazil (CNPQ) has been developed with similar objectives.

Remote sensing and GIS are principal tools aimed to enhance basic socio and environmental knowledge about Amazonian coast. Maps were prepared in strategic and tactical scales through the use of digital elevation models derived from the SRTM (Shuttle Radar Topographic Mission) and optical sensors data (Cohen & Lara, 2003; Souza Filho & El Robrini, 2000; Souza Filho, 2005; Szlafsztein & Sterr, 2007; Silva et al., 2009), synthetic aperture radar (SAR) data (Souza Filho & Paradella, 2002 and 2005; Costa 2004; Souza Filho et al., 2005b; Silva et al., 2009), and the combination of some of them (Souza Filho & Paradella, 2005; Gonçalves et al., 2009; Rodrigues & Souza Filho, 2011).

Oil spill environmental sensitivity maps, adapted to the peculiarities of the Amazonian region (Souza Filho et al., 2004) were drawn at tactical scales through the use of Radarsat and Landsat sensors (Gonçalves et al., 2009; Teixeira & Souza Filho 2009; Boulhosa & Souza Filho, 2009), and operational scale through the use of High Resolution remote sensing (Andrade et al., 2010; Rodrigues & Szlafsztein 2010; Andrade et al., 2009). Over the past decade were reached advances in identification and assessment of sensitivity through spatial maps, the impacts to oil spill analyses, and oil spill risk in Amazonian coastal. The goal of this book chapter is to present a review of the oil spills environmental sensitivity mapping activities using remote sensing and GIS tools in the Amazonian coastal zone of Brazil.

## 2. Remote sensing and coastal environmental sensitivity for oil spill

### 2.1 Remote sensing

Remote sensing tools are essential for the construction of maps. These tools help in the precise delimitation of coastlines and specific landforms. The selection of appropriate remote sensing data and applicable digital image processing techniques involves a compromise between costs and mapping capabilities, including coverage area, and spatial resolution (Green, 2000).

For risk maps, remote sensing are fundamental. Risk appears in a broader context in humans  transform of the natural into a cultural environment, with the aim of improving living conditions and serving human wants and needs (Turner et al., 1990). There are several sources of hazards to the environment and to society, some of them originated in human activities (Smith & Petley, 2008).

Oil spills are an example of this technological risk. Information and detection about oil spills can be collected through remote sensing tools for prevention planning, as well as river/ocean pollution monitoring and restoration. Some reviews of the use of remote sensing and oil spills including Brekke & Solberg (2005) and Fingas & Brown (2000).

---

[1] PETROBRAS is the large oil company in Brazil

For a coastal environment, remote sensors can provide information about the physical characteristics of the shoreline, coastal ecosystems dynamics, water quality, and land use/occupation. This information could be mapped at different scales generating cartographic products using all types of sensors and specific digital image processing (Jensen, 1996). Sensors can provide timely and valuable information about oil spills, including the location and extent, thickness distribution, and oil type in order to estimate environmental damage, take appropriate response activities, and to assist in clean-up operations for oil spill contingency planning (Grüner, 1991).

The most common sensors utilized to detect oil spills and to map coastal environments are: optical (visible, infrared sensors and ultraviolet sensors) or radar. Both types of sensor may be acquired at terrestrial, sub-orbital or orbital levels. At terrestrial level, both still and video cameras are commonly used. At the sub-orbital level (or airborne remote sensing), airplanes is the most commonly utilized platform. At the orbital level, satellites are usually used as a platform for sensors. Satellite differs from airborne remote sensing due to timing and frequency of the data collection, the demand of good climate conditions and the time required for processing the dataset (Jha et al., 2008). Aiming to compare sensors, a brief description is given in Table 1.

### 2.1.1 Optical

Optical sensors can be composed by three bands in the electromagnetic spectrum. These sensors are usually composed by multispectral bands in visible and infrared intervals from the electromagnetic spectrum. In the visible region (350 to 750 nm), oil has a higher surface reflectance than water, but also shows limited nonspecific absorption tendencies (Jha et al., 2008). Instruments such as cameras, films and spectrometers are optical techniques for remote sensing with the benefit of low cost. Normally, visible sensors cannot operate at night as they depend on the reflectance of sunlight, but, in the case of oil spills they can be used to create environmental and logistic maps of the coast to subsidize field trips and first risk management decisions. The infrared sensors are at the 0,7-14 μm intervals in the electromagnetic spectrum. Solar radiation is partially absorbed and emitted as thermal energy by oil. This is thermal energy concentrated in the thermal infrared region with a distinct spectral signature; water has a higher emissivity (Salisbury et al., 1993). Infrared sensors can provide information about the relative thickness of oil slicks, but these sensors are unable to detect emulsions of oil in water when oil is diluted to 70% water (Fingas & Brown, 1997). Infrared is reasonably inexpensive, but has limitations related to false positive results generated by weeds and shorelines (Fingas & Brown, 2000).

Ultraviolet sensor scanners capture ultraviolet radiation (0,003 – 0,38 μm) reflected by the sea surface for detecting oil spills. Oil is more reflective than water in the ultraviolet region. Limitations of this sensor are related to undetected information greater than 10 microns and false images produced by such hindrances as wind slicks, sun glints, and biogenic material (Grüner, 1991).

### 2.1.2 Radar

Radar is an active sensor (not dependent on electromagnetic radiation from the sun) and operates in a radio wave region (1m – $10^4$m). Radar sensors can have two principal

| Sensors | | Platform | Spatial resolution (m) | Over-pass Frequency (days) | Imagery area | Application |
|---|---|---|---|---|---|---|
| RADAR | SAR | Spaceborne — ERS-2 | 30 | 3, 35 and 176 days | 100km | Identify large offshore spills and coastal environments – Strategic planning and monitoring |
| | SAR | Spaceborne — RADARSAT-1 | 8-100 | 24 days | 45-500km | |
| | SLAR | Airborne — Airplane | 10-50 | As required | 60-80km | Detect and identify the polluter, the extent and type of oil spill and the cleaning necessity; Environmental mapping – Strategic and tactical planning |
| | SAR | Airborne — Airplane | 1-10 | As required | - | |
| OPTICAL | MSS,TM, ETM, ETM + | Spaceborne — Landsat 5 Landsat 6 Landsat 7 | 15-120 | 16 days | 183-185km | Detect oil spill if the weather conditions are good; can discriminate false positives; identify and mapping environments – Strategic and tactical planning |
| | HRV | Spaceborne — Spot-2 Spot-3 | 10-20 | 26 days | 60x60km /100km | |
| | CCD | Spaceborne — Cbers-1; Cbers-2 | 20 | 26 days | 113 km | Detect oil spill if the weather conditions are good; identify and mapping environments – Strategic and tactical planning |
| | IRMSS | Spaceborne — Cbers-1; Cbers-2 | 80-160 | 26 days | 120 km | Detect oil spill if the weather conditions are good; capable to detect thermal surface differentiations - Strategic and tactical planning |
| | WFI | Spaceborne — Cbers-1; Cbers-2 | 260 | 5 days | 890 km | Detect oil spill if the weather conditions are good; monitoring; identify and mapping environments – Strategic planning |
| | Video camera | Airborne — Airplane | Altitude Dependent | As required | - | Oil spill and coastal environmental documentation. The infrared sensor for measure the thickness of oil slicks – Operational planning |
| | Still camera | Airborne — Airplane | Altitude Dependent | As required | - | |

Table 1. Characteristics of some existing sensors for oil spill management applications.

instruments: Side-Looking Airborne Radar (SLAR) and Synthetic Aperture Radar (SAR). Radar is a very powerful and useful sensor for searching large areas, observing oceans at night, and capturing images during cloudy weather conditions. The presence of an oil spill can be detected without thickness estimation or oil type recognition. In the radar image, the leak appears as a dark area in contrast to the bright image of the ocean because radar waves are reflected by capillary waves on the ocean (Brown et al., 2003). For a coastal environment, mapping SAR is already considered to be a powerful tool for geomorphologic mapping, providing relevant information about the emergence and submergence of the coast (Souza Filho et al., 2009a).

SLAR is an old technology predominantly used for airborne remote sensing (Fingas & Brown, 2000). Airborne surveillance is limited by high costs and is less efficient for wide area observation due to its limited coverage. SAR has greater spatial range and resolution than the SLAR because it uses the forward motion of the aircraft to synthesize a very long antenna, thereby achieving very good spatial resolution, at the expense of sophisticated electronic processing (Mastin et al., 1994). SAR can be used to provide an initial warning because aircrafts are more suitable to identify the polluter, the extent, and the type of spill.

For large scale oil spill detection, satellite platforms, including ERS-1 and -2, Radarsat, and JERS-1, are commonly used for large scales oil spills (Fingas & Brown, 2005). Radar satellites, including ERS-1 and -2, Radarsat, and JERS-1, have been useful for mapping known large offshore spills (Biegert et al., 1997). On the other hand, optical satellite imagery does not offer much potential for oil spill detection (Fingas & Brown, 2000). However, to map coastal environments, geomorphology and its sensitivity, multi-sensor data fusion such as optical and radar has proved to be a successful tool (Souza Filho et al., 2009b).

## 2.2 Coastal environmental sensitivity to oil spills

The inter-relationships involving natural resources and human societies have led to a concentration of human activities, services and survival strategies in the coastal environment (Viles & Spencer, 1995; Muehe & Neves, 1995; Pernetta & Elder, 1992). The unique natural geodynamics, the highly productive and extremely diverse biological systems extending from coastal lands to deep water regions (Malthus & Mumby, 2003), the growing land use changes and the pressure on natural resources (MEA, 2005) transform the coastal zone into a conflict area. Oil exploration, transportation and storage have increased the technological risk in this zone.

Areas neighboring major ports (environmental and human populated) may be affected by oil transportation, tank cleaning and oil storage procedures in a port area (Noernberg & Lana, 2002). One of the initial concerns about oil spills result in a necessity for the construction of maps that indicate which type of environment and human resources will be affected. In the mid 1970s, scientists from the National Oceanic and Atmospheric Administration (NOAA) and the American Coast Guard of the United States began to study and classify the sensitivity of coastal environments to oil spill.

This classification was based, initially, on the vulnerability index to oil spills proposed by Gundlach & Hayes (1978). Coastal area is segmented considering environmental and geomorphologic characteristics and then classified using the Vulnerability Index, scaled from 1 (low) to 10 (high). This Vulnerability Index became the standard for coastal management, planning and research about the effects of oil spills on different types of coastline. Over time, this index evolved and was modified, leading eventually to the development of the Environmental Sensitivity Index (ESI).

The ESI should be represented cartographically as maps in different scales for different goals. The first ESI map was produced in 1979, in response to the advance toward the coast of oil resulting from the blowout of the IXTOC 1 oil-well in the Gulf of Mexico. In the 1980s, ARPEL produced an innovative ESI atlas for the whole coast of the United States, including Alaska and the Great Lakes, to be used for the planning of contingency measures in response to oil spills (NOAA, 2002). From this moment on, ESI maps have been an integral component of response and contingency planning for oil spills, looking for the protection of life, the reduction of environmental impacts, and facilitation of the response efforts. These atlases were integrated by color printed maps on a two dimensional representation of a three-dimensional world and high production costs.

After the 1990's, NOAA (2002) standardized output formats and symbols for ESI maps construction. The basic necessary information is 1) shoreline classification; 2) biological resources; 3) human-use resources. The shoreline classification scheme is based on an understanding of the physical and biological characteristics of the shoreline environment. Relationships among physical processes, exposure to wave and tidal energy, slope, substrate type (i.e. grain size, mobility, penetration and/or burial, and mobility), and associated biota produce specific geomorphic/ecologic shoreline types. Shoreline classification helps to identify oil spill origin and impacts and the best cleanup methods for a specific shoreline type. The sensitivity ranking was developed for the estuarine settings and is slightly modified for lakes and rivers. The human use resources relate to specific, valuable specific areas because of their use, such as beaches, parks and protected marine areas, water intakes, fisheries, tourism, economic sectors, and archaeological sites. The biological resources include the study and maps of oil-sensitive biological and ecological resources.

## 3. Remote sensing and coastal environmental sensitivity in Brazil

Brazil has an expansive coastline through the equatorial region to the subtropical latitude of the south hemisphere. The length is approximately 8.500 km with 17 of the 26 states of the country lying on the coast of the Atlantic Ocean. The Brazilian coast is defined by the National Plan for Coastal Management (law 7661/1988), as the geographic space where there are air, sea and land interacts, which includes renewable and non-renewable resources along a maritime and terrestrial border.

A diversity of coastal environments and population densities are found along the Brazilian coast. Population is higher in state capitals than in the other coastal municipalities. Environments vary from very productive, such as mangroves, to rocky and artificial man-made structures. Man-made structures, such as ports are established along the entire coastline of Brazil (Figure 2).

Ports are high-risk areas, and oil spill monitoring is clearly important there. In 2000, two large oil spills occurred at Baía de Guanabara (Rio de Janeiro) and Paraná, both resulting from pipeline ruptures. After these accidents, fundamental changes have been made to

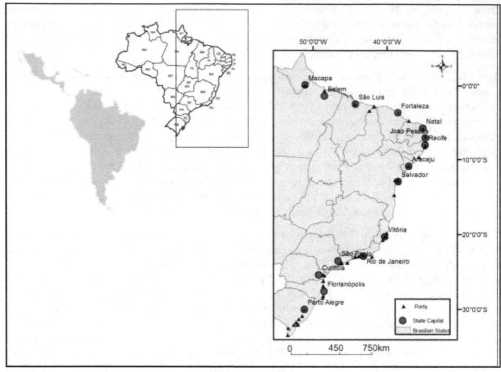

Fig. 2. Main ports of the Brazilian coast (IBAMA, 2011).

the environmental policies of Brazil and PETROBRAS (Souza Filho et al., 2009b), in order to give priority to prevention and mitigation activities.

After these accidents, the prevention and mitigation oil spill impacts became a priority. The law 9966/2000 was established to regulate the activities of prevention, control and supervision of pollution caused by oil and other dangerous substances in Brazilian waters. The Ministry of the Environment has the responsibility of identifying, locating and defining the boundaries of the ecologically sensitive areas to oil spills. Ecologically sensitive areas are defined as regions where special attention is needed in order to protect and preserve the environment from pollution by regulatory and preventive measures (MMA, 2002). In this context, Araújo et al. (2002) published the document "Basic Manual for the Elaboration of Maps of Environmental Sensitivity to Oil Spills in the Petrobras System: Coastal and Estuarine Environments" edited by the Ministry of the Environment (Climate Change and Quality Environmental Secretary) and based on PETROBRAS (2002) and NOAA (2002).

This change in attitude was reflected in a large production of oil spill environmental sensitivity maps for the Brazilian coast (Araújo et al., 2006) (Table 2). These maps are produced in order to support environmental management and the elaboration and implementation of contingency plans. The ESI preparation was intensified through Decrees 4136/2002 and 4871/2003.

| Author | Spatial Resolution | Sensor | Method | Scale | Case study (State) |
|---|---|---|---|---|---|
| Romero, et al. (2010) | High | Aerial photography | Literature review, Field data collection | Operational | Cananéia-Iguape Estuary (São Paulo) |
| Silva et al. (2010) | Low | SRTM, Radarsat | Highly precision field data collection with DGPS, Previous database research, visual classification | Tactical, Operational | Potiguar Sedimentary Basin (Rio Grande do Norte) |
| | Moderate | Landsat 7 ETM+, Cbers-2, Cbers-2B | | | |
| Cantagalo et al. (2008) | High | Aerial photography | Visual classification, field data collection | Tactical, Operational | Santos Estuary (São Paulo) |
| Carvalho & Gherardi (2008) | Moderate | Landsat 7 ETM | Automatic classification, visual interpretation, Field data collection | Tactical, Operational | Potiguar Sedimentary Basin (Ceará and Rio Grande do Norte) |
| Bellotto & Sarolli (2008) | Moderate | Landsat 7 ETM+ | Visual interpretation, Previous database research, Field data collection | Operational | Municipality of Imbatuba (Santa Catarina) |
| Poletto & Batista (2008) | Moderate | Cbers | Visual interpretation, Previous database research, Field data collection | Tactical, Operational | Municipality of Ubatuba (São Paulo) |
| | High | Aerial photography | | | |
| Rocha-Oliveira et al. (2008) | Moderate | Landsat 7 TM+ | Visual interpretation, Field data collection | Operational | Southeast and south area (Santa Catarina) |
| Silva et al. (2008) | Moderate | Landsat 7 TM+ | Visual interpretation, field works, literature review | Operational | Santa Catarina Island and surrounding areas (Santa Catarina) |
| Araújo et al. (2007) | Moderate | Landsat 7 ETM+ | Visual interpretation, Field data collection | Operational | Municipalities of Itapoa, Barra Vellha, Piçarras, Itajaí, Balneário Camboriú (Santa Catarina) |
| | High | Aerial photography | | | |
| Chacaltana (2007) | High | Ikonos | Visual interpretation, Field data collection | Operational | Vitória Bay (Espírito Santo) |
| Lima et al. (2008) | High | Aerial photography | Visual interpretation, Field data collection | Operational | São Sebastião Island (São Paulo) |
| Wieczorek et al. (2007) | High | Aerial photography | Visual interpretation, Field data collection | Operational | Cananéia-Iguape Estuary (São Paulo) |

| Author | Spatial Resolution | Sensor | Method | Scale | Case study (State) |
|---|---|---|---|---|---|
| Castro et al. (2006) | Moderate | Landsat 5 TM, Landsat 7 ETM+ | Database development, geomorphology; hydrodynamic, waves energy, currents direction; slope and grain size of profile beach | Operational | São Bento, Galinho Municipalities (Rio Grande do Norte) |
| | High | Aerial photography | | | |
| Souto et al. (2006) | Moderate | Landsat 5 TM, Landsat 7 ETM+ | Normalized Difference Vegetation Index, Automatic classification, Visual interpretation, Field data collection | Operational | Ponta Macau (Rio Grande do Norte) |
| | High | Aerial photography | | | |
| Souza, et al. (2005) | Moderate | Landsat 5 TM, Landsat 7 ETM+, SPOT, Cbers-2 | Database utilization; visual interpretation, Field data collection | Tactical | Northern coast (Rio Grande do Norte) |
| | High | Ikonos | | | |
| Noernberg & Lana (2002) | Moderate | Landsat TM | Database access, digital processing | Operational | Paranaguá Estuary (Santa Catarina) |

Table 2. Principal studies of oil spill coastal sensitivity using remote sensing techniques in order to generate ESI maps in Brazil organized by date (Amazon Region are not included).

In Brazil, the ESI maps were also developed in a cartographic plan at strategic, tactical and operational scale for the role country. As an initial step, the tools of remote sensing and GIS are necessary to ESI maps construction and to comprehend differential spread of the technological risk for the country's coasts. Mostly moderate and high resolution images were used to produce these maps.

Moderate resolution images (e.g. RADARSAT-1 and Landsat TM/ETM/ETM+) and SRTM derived digital elevation models have been used to map the Brazilian coastal zone at strategic and tactical scales. Studying an oil spill emergency due to a pipeline rupture in Guanabara Bay (Rio de Janeiro), Bentz and Miranda (2001) found that RADARSAT-1 provided suitable temporal coverage. Once cloud cover, haze and the eight-day revisit schedule (using both Landsat-5 and -7) prevented Landsat from being used systematically for oil spill monitoring. In the same case Thematic Mapper (TM) sensor was used to capture images after the oil spill emergency where a pipeline ruptured (Bentz & Miranda, 2001).

Carvalho & Gherardi (2008) used Landsat 7 ETM+ images to generate land use and land cover maps, as well as ESI maps in Northeast Brazil, aiming for oil spill contingency planning and emergency responses. A fusion of multispectral and panchromatic ETM images via IHS (Intensity-Hue-Saturation) transformation was used. Then socioeconomic information was inserted using automated and visual image interpretation.

High resolution images have mostly been used for operational ESI maps production in the states of São Paulo and Rio Grande do Norte using aerial photography and Ikonos. Visual interpretation, together with field data collection, has been the principal methodological procedure. Most areas have mangroves, conservation units and are surrounding by intensive technological activities.

The methodology, standards and technical specifications for determining coastal sensitivity follow Araújo et al. (2002). The principal steps for shoreline identification are: 1) Analysis of the available literature, aerial photographs, maps of the entire area; 2) Aerial reconnaissance of the entire area and selection of detailed study areas; 3) Mapping of major features in representative areas; 4) Collection of sediment from the intertidal zone and biologic floral and faunal groups samples; construction of beach topographic profile; 6) Analysis of the sediment sample sizes; 7) Data compilation and classification; and 8) Construction of detailed sensitivity maps.

Colors are used indicate the ESI and symbols to the human and biological resources. Each number is represented by a color index. Two environments may occur at the same coastal segment; in that case, both colors of the separated lines should be displayed, one inside and the other outside. In the case of intertidal zones, for example, the intertidal plain should display colors according to the differences of sediment sizes to the high tidal line and the low tidal line. Table 3 compares ESI specification defined by the Ministry of the Environment (MMA, 2002) with the original defined by NOAA (2002).

| ESI number | Color | NOAA (2002) | MMA (2002) |
|---|---|---|---|
| 1 | | Exposed rocky shores and man-made structures; rocky cliffs with boulder talus base | Exposed rocky shores; exposed rocky sedimentary cliffs; exposed solid man-made structures |
| 2 | | Exposed wave-cut platforms in bedrock, scarps and steep slopes in clay | Exposed medium to high declivity rocky shores; exposed sandy substrate with medium declivity |
| 3 | | Fine to medium grained sand beaches; Scarps and steep slopes in sand; Tundra cliffs | Fine to medium grained sand in dissipative beaches; continuous and multiple beach strings; Scarps and steep slopes in sand; exposed dune field |
| 4 | | Coarse-grained sand beaches | Coarse-grained sand beaches; exposed; exposed fine to medium grained sand intermediary beaches; sheltered fine- to medium- grained sand beaches |
| 5 | | Mixed sand and gravel beaches | Mixed sand and gravel beaches, coral reefs fragments; vegetated abrasion platform; sandy reefs |

| ESI number | Color | NOAA (2002) | MMA (2002) |
|------------|-------|-------------|------------|
| 6 |  | Gravel beaches; Riprap gravel beaches (cobbles and boulders) | Gravel beaches; dendritic limestone coast; platform with lateritic concretion |
| 7 |  | Exposed tidal flats | Exposed sandy tidal flats; low tide platform |
| 8 |  | Sheltered: scarps in bedrock, mud or clay, rocky shores (impermeable/permeable), solid man-made structures, riprap, rocky rubble shores; Peat shoreline | Sheltered scarps in bedrock (permeable and non permeable); Scarps and steep slopes in sand; permeable sheltered man-made structures (riprap) |
| 9 |  | Sheltered tidal flats; Vegetated low banks; hypersaline tidal flats | Sand tidal flats; sheltered mud tidal flats; coral reefs |
| 10 |  | Salt and brackish water marshes; Freshwater marshes; Swamps; Scrub-shrub wetlands: mangroves; Inundated low-lying tundra | Delta and vegetated sand bars; sheltered wetlands; salt marshes saline wetlands; mangroves |

Table 3. ESI comparison between NOAA (2002) and Ministry of the Environment (MMA, 2002) classification.

## 4. Remote sensing and coastal environmental sensitivity in Amazon

The coastal zone of the Brazilian Amazon is composed by tree states: Amapá, Pará and Maranhão. According to the IBGE (2011), the total population of Amapá State is 669,526 distributed in 16 municipalities; the state capital, Macapá, concentrates 59% of this population. The state of Pará has a total population of 7,581,051 distributed in 143 municipalities; Belém comprises 18%. Maranhão state has a total population of 6,574,789 distributed in 217 municipalities; São Luis comprises 15% of this population. The population density in capital cities is over 100 hab/km², while in other coastal municipalities vary from 10 to 50 hab/km².

Until 21st century most of the coastal zone of the north of Brazil had sectors virtually devoid of information, or where data was available, it was non-systematized and both temporally and spatially non-continuous. The most important environmental dataset is related to the large continuous and well-developed mangroves - *Rhizophora mangle, Avicennia germinans* and *Laguncularia racemosa* (Szlafsztein, 2000). The mangroves have ecological and socioeconomic importance due to communities' livelihoods, and they are considered a protected ecosystem. The main activities are fishing, collecting shrimp and crabs (Andrade et al., 2010; Andrade et al., 2009).

However, port complexes and industries have been established alongside residential, protected areas and fishing grounds. For example, in Piatam Mar context the principal ports chosen to develop oil mapping were "Santana" (State of Amapá); "Itaqui" (State of Maranhão); "Outeiro", "Miramar" and "Vila do Conde" (State of Pará). The biological information was intensively identified in "Lago Piratuba biological reserve" (Amapá); "Soure extractive reserve" (Pará) and "Ilha dos Caranguejos Environmental Protection Area" (Maranhão). According to Souza Filho et al., (2009a) these conservation units work as control areas, given both their well-preserved conditions and their proximity to transportation routes due to proximity to protected areas along the ports mentioned above (Figure 3).

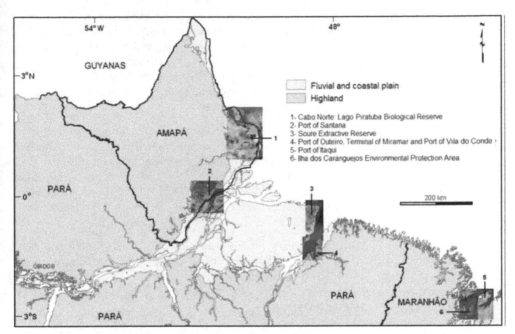

Fig. 3. Principal ports and environmental protected areas in the Amazon coast (source: Souza Filho et al., 2009a).

Oil spills are a potential risk around these port areas which can affect the environment, human population infrastructure and livelihood, resulting from the transportation process, as well as tank cleaning and oil storage procedures within the area of the port (Noernberg & Lana, 2002). To comprehend the oil impact, it is necessary to analyze the coastal Amazonian environment as a whole. PETROBRAS established and financed nine projects to deal with this subject, among them, the "Environmental Sensitivity Map to Oil Spill in Guajará Bay (PA)" (2001 – 2003), the "JERS-1, RADARSAT-1 and ALOS/PALSAR application in monitoring and mapping Amazon coastal environments: an approach for multi-temporal environmental sensitivity maps to oil spill" (2004 - 2006), PIATAM MAR (2004 - 2010) and currently "Elaboration of ESI maps for Pará-Maranhão and Barreirinhas basin" (2012 until 2014).

The PIATAM MAR project was implemented in Northern Brazil and was led by the Federal University of Pará, the Federal University of Rio de Janeiro and PETROBRAS. The general aims proposed are: the consolidation of a multidisciplinary researcher's network that are active in the Amazonian coastal zone; the development of technological tools and infrastructure to support local monitoring and environmental management; and ESI maps construction (Souza Filho et al., 2009a).

Initially, the researchers of PIATAM MAR project compiled environmental data and other information available on the Amazon coastal zone. The results are integrated in the book, "Bibliography of the Amazon Coastal Zone: Brazil" (Souza Filho et al., 2005a). Meanwhile, a computational database system using the MYSQL language was developed and used as a basis for the development of a geographic information system called SIGmar.

All of these initial steps support the subsequent aim of PIATAM MAR: the construction of ESI maps. From 2006 to 2010 socio-economic and environmental data were integrated in ESI maps. These maps have been developed through the SIGmar and the extensive use of remote sensing. ESI maps create an operational alternative for the monitoring and mapping of the Amazonian coastal zone and provide guidelines for the use of the InfoPAE (Computerized Emergency Action Plan Support) System on the Amazonian coast (Souza Filho et al., 2009a).

Two considerations should be taken into account when mapping and monitoring oil spills in the Amazonian coastal environments. First, the unique complex environmental dynamics of the Amazon basin have demanded an adaptation of ESI classification with values from 1 (low) to 10 (high) sensitivity (Souza Filho et al., 2004) (Table 4). Second, the Amazonian coast is situated in the intertropical convergence zone (ITCZ) that is located near the equator and has a broad area of low atmospheric pressure. Therefore, there is a huge cloud cover between December and May which limits the use of some kinds of sensors.

Coastal ESI mapping for the Amazon uses remote sensing as an indispensable and very powerful tool. Oil spill and environmental sensitivity to oil spills in the Amazon were mostly mapped during the PIATAM MAR project. Table 5 shows the most important scientific results in this study area. The perspectives of the ESI adaptation proposed by Souza Filho et al. (2004) were extensively used.

The whole Amazon coastal zone was mapped with spatial resolution of 90 m based on the processing and images mosaics of SRTM images and 30 m of RADARSAT-1 Wide 1 images and mosaics of JERS-1 SAR. This sensor was chosen given the six months of unfavorable climatic conditions; radar sensors (Synthetic Aperture Radar – SAR) are used for strategic scale.

On a tactical scale, multi-sensor data fusion between microwaves sensors and optical sensors are considered to be the most important source of spatial data for geomorphologic recognition and basic coastline characteristics. The commons sensors fusion are made in general with low resolution data from RADARSAT-1 Wide 1 and JERS-1 SAR mosaic, together with moderate spatial resolution data (10–30 km) from Landsat series (MSS, TM and ETM+) and Cbers-2 images (20m). The multi-fusion of optical (Landsat 5 TM) and radar (RADARSAT-1) sensors had a particular emphasis on the evaluation of the new hybrid sensor product combining PCA (Principal Component Analysis) and IHS components. In areas with little or no data, this fusion method from multi sensors to orbital images, together with field data are economically efficient and provide a good environment sensitivity characterization (Rodrigues & Souza Filho, 2011).

Hydrological dynamics with flood area delimitation could be differentiated by the use of JERS-1, L band (Santos et al., 2009), which is important in a region dominated by different tidal regimes that can amplify the area affected by oil spills. Methods include visual and automatic classification leading to good results in identifying widespread occurrence of

| ESI | Amazon Coastal Environment |
|-----|----------------------------|
| 1A | Exposed rocky shores |
| 1B | Exposed, solid man-made structures |
| 1C | Exposed rocky cliffs with boulder talus base |
| 2 | Exposed scarps and steep slopes in clay |
| 3A | Fine to medium grained sand beaches |
| 3B | Scarps and steep slopes in sand |
| 4 | Coarse-grained sand beaches |
| 5 | Mixed sand and gravel banks and beaches |
| 6 | Riprap |
| 7 | Exposed tidal flats |
| 8A | Sheltered scarps in bedrock, mud, or clay |
| 8B | Sheltered, solid man-made structures |
| 8C | Sheltered riprap |
| 8D | Peat shorelines |
| 9A | Sheltered tidal flats |
| 9B | Vegetated low banks |
| 9C | Hypersaline tidal flats |
| 10A | Salt, and brackish-water marshes |
| 10B | Freshwater marches, aquatic vegetation |
| 10C | Intertidal mangrove |
| 10D | Supratidal mangrove |

Table 4. ESI shoreline classification for the Amazon Coast, modified by Souza Filho et al. (2004) based on the proposals of NOAA (2002) and Araújo et al. (2002).

flooded mangrove forests. This environment is considered to be the most oil-sensitive habitat described in Table 4 - ESI Ranking specification = 10c and 10d (Souza Filho et al., 2005a).

High resolution images, such as Ikonos, were used for operational scale mapping. The resolution of 1 m provides a detailed geomorphic map, and it's also possible to map the potentially hazardous industrial structures stratified by type of hazard. However, the use of Ikonos images is limited when cloud cover is higher than 25%. As a result the images are mostly inadequate between March and June (Andrade et al., 2010). On the other hand, biological and socioeconomic resources, risk areas and oil spill hazard zones of storage and platform transportation can be better identified and delimitated (Rodrigues & Szlafsztein, 2010; Andrade et al., 2009). This location contributes to planning and management strategies and cleaning efforts.

| Author | Map Type | Spatial Resolution | Sensor | Method | Scale | Study case |
|---|---|---|---|---|---|---|
| Andrade et al. (2010) | Oil spill Vulnerability | High | Ikonos | Visual interpretation, Field data collection | Operational | Municipality of São Luis (Maranhão) |
| Rodrigues & Szlafsztein (2010) | Oil spill risk | High | Ikonos | Visual interpretation, Field data collection | Operational | Municipality Barcarena (Pará) |
| Andrade et al. (2009) | Oil spill hazard representation and susceptible socioeconomic resources | High | Ikonos | Visual interpretation, Field data collection | Operational | Municipality of São Luis (Maranhão) |
| Boulhosa & Mendes (2009) | ESI | Moderate | Spot-5 | Visual interpretation, Field data collection | Operational | Municipality of Barcarena (Pará) |
| Boulhosa & Souza Filho (2009) | ESI | Moderate | Landsat 7 ETM+, SRTM High, aerial photography | Automatic classification, multi-fusion sensors, Field data collection | Tactical | Municipalities of Maracanã, Santarém Novo, Salinópolis, Cuiarana, São João de PIrabas, Santa Luzia and Primavera (Pará) |
| Gonçalves et al. (2009) | ESI | Moderate | Landsat 7 ETM, Radarsat 1 | Automatic classification, multi-sensor fusion, field data collection | Tactical | Municipality of Belém (Pará) |
| Souza Filho et al. (2009) | ESI | Moderate | Landsat 5 TM, Radarsat 1 | Visual classification | Tactical | Municipality of Bragança (Pará) |
| Novaes et al. (2007) | ESI | Moderate | Landsat 5 TM | Visual interpretation, Field data collection | Operational | Municipality of São Luis, (Maranhão) |

Table 5. Results published in a scientific paper related to remote sensing use and sensitivity environment to oil spill in Amazon coast.

## 5. Conclusions

Remote sensing techniques are used for risk identification, assessment and analysis. The technological risk of oil spills needs continuous planning and monitoring actions. The availability of airborne and satellite remote sensing provides a diversity of resolution and sensors required to construct environmental sensitivity maps, using basic information about socioeconomic and biological resources and geomorphic characteristics.

Remote sensing and ground confirmation provide accurate information about this basic information. In particular, the coastline is usually mapped in detail with both optical and radar sensors. The multi-sensor data fusion of an optical moderate sensor with radar has been extensively used in the Amazon region to provide basic information about coastal environments. Radar is a very powerful tool once it can operate in difficult weather conditions. It provides detailed information about shoreline irregularities and geomorphic units if the texture and the altitude of this type of images are precise. Optical sensors are used for environmental differentiation once land cover and land use have different spectral responses.

Studies in Brazil regarding oil spills have increased after 2000, and ESI maps have been generated at different scales for different areas along the coast. Remote sensing tools were essential to achieve initial and advanced cartographic information in a context of the diversity of the environment, information and cartographic background. Particularly in the Amazon, little background information about the coastline existed before the PIATAM MAR project. In the context of this project, the Amazon coast was previewed at strategic scale with the use of a moderate sensor. ESI maps were produced at the tactical and operational scales and it was possible to map the coastal environment and organize information about socioeconomic and biological resources. A large, extensive mangrove system coexists with industrial port areas on the Amazon coast with a high sensitivity to oil spills, which should to be continuously monitored with remote sensing techniques.

## 6. References

Andrade, M.; Souza Filho, P. & Szlafsztein, C. (2009). High Resolution Images for Recognition of the Susceptibility of Social Economic Resources to Oil Spill in the Itaqui-Bacanga Port Complex, Maranhão, Brazil. *Journal of Integrated Coastal Zone Management*, vol. 9, pp. 127-133.

Andrade, M; Szlafsztein, C., Souza Filho, P., Araújo, A. & Gomes, M. (2010). A socioeconomic and natural vulnerability index for oil spills in an Amazonian harbor: A case study using GIS and remote sensing. *Journal of Environmental Management*, vol 91, pp. 1972-1980.

Araújo, S., Silva G. & Muehe, D. (2002). Manual básico para elaboração de mapas de sensibilidade ambiental a derrames de óleo no sistema Petrobras: Ambientes Costeiros e Estuarinos. CENPES/Petrobras, Rio de Janeiro, 134 pp.

Araújo, S.; Silva, G. & Muehe, D. (2006). *Mapas de sensibilidade ambiental a derrames de óleo - Ambiente costeiros, estuarinos e fluviais.* Rio de Janeiro Rio de Janeiro. CENPES/Petrobras, pp.168 Ed. 2ª, ISBN: 8599891014

Araújo, R.; Petermann, R.; Klein, A.; Menezes, J.; Sperb, R. & Gherardi, D. (2007). Determinação do ìnndice de sensibilidade do litoral (ISL) ao derramamento de óleo para regiões norte e centro-norte da costa de Santa Catarina (SC), *Gravel*, N°5, pp. 45-73.

Belloto, V. & Sarolli, V. (2008). Mapeamento da sensibilidade ambiental ao derramamento de óleo e ações de resposta para a região costeira e área portuária de Imbituba, SC, Brasil. *Brazilian Journal for Aquatic Sciences and Technology*, vol. 12, N°2, pp. 115-125.

Bentz, C. & Miranda, F. Application of remote sensing data for oil spill monitoring in the Guanabara Bay, Rio de Janeiro, Brazil. In: *Proceedings of the International Geosciences and Remote Sensing Symposium (IGARSS)*, Sydney, 2001.

Berredo J.; Mendes A.; Sales M. & Sarmento J. (2001). Nível de Contaminação por óleo nos sedimentos de Fundo e na água do Rio Pará, decorrente do acidente com a balsa Miss Rondônia. In: Prost M. & Mendes A. (Org.). *Ecossistemas Costeiros: Impactos e Gestão Ambiental*. Museu Paraense Emílio Goeldi, Belém, pp. 153 - 165.

Biegert, E., Baker, R., Berry, J., Mott, S. & Scantland, S. (1997). Gulf Offshore Satellite Applications Project Detects Oil Slicks Using Radarsat, In: *Proceedings of International Symposium: Geomatics in the Era of Radarsat*, Ottawa, Canada.

Boulhosa, M. & Mendes, A. (2009). Mapeamento dos Índices de Sensibilidade Ambiental ao derramamento de óleo através de Imagens SPOT 5, na região portuária de vila do conde - Barcarena - PA. *XIV Simpósio Brasileiro de Sensoriamento Remoto*, Natal, Brasil, pp. 3597-3603.

Boulhosa, M. & Souza Filho, P. (2009). Reconhecimento e mapeamento de ambientes costeiros para geração de mapas de índice de sensibilidade ambiental ao derramamento de óleo, Amazônia Oriental. *Revista Brasileira de Geofísica*, vol. 27, pp. 23-37.

Brekke, C. & Solberg, A. (2005). Oil spill detection by satellite remote sensing. *Remote Sensing of Environment*, vol. 95, pp. 1–13.

Brown, C. & Fingas, M. (2003). Development of airborne oil thickness measurements. *Marine Pollution Bulletin*, vol. 47, pp. 485 - 492.

Cantagallo, C.; Milanelli, J. & Dias-Brito, D. (2007). Limpeza de Ambientes Costeiros Brasileiros Contaminados pó Petróleo: uma revisão. *Pan-American Journal of Aquatic Sciences*, vol 2, pp. 1-12.

Castro, A.; Grigio, A.; Souto, M.; Amaro, V. & Vital, H. (2006). Modeling and development of a geographic database: application to the elaboration of oil-spill environmental sensitivity maps in coastal areas of the Rio Grande do Norte state. *Journal of Coastal Research*, vol. 39, pp. 1436-1440.

Carvalho, M. & Gherardi, D. (2008). Mapping the environmental sensitivity to oil spill and land use/land cover using spectrally transformed Landsat 7 ETM data. *Brazilian Journal for Aquatic Sciences and Technology*, vol. 12, N° 2, pp. 1-9.

Chacaltana, T. (2007). Mapeamento de áreas de sensibilidade ambiental ao derrame de óleo na Baía de Vitória, ES. *Master dissertation*. Federal University of Espírito Santo. p.140.

Cohen, M. & Lara R. (2003). Temporal changes of mangrove vegetation boundaries in Amazonia: Application of GIS and remote sensing techniques. *Wetlands Ecology and Management*, vol 11, pp. 223 -231.

Costa, M. (2004). Use of SAR satellite for mapping zonation of vegetation communities in the Amazon floodplain (10): 1817-1835. *International Journal of Remote Sensing*, vol 25, pp. 1817-1835.

Dutrieux, E.; Canovas, S.; Denis, J.; Hénocque, Y.; Quod, J.; Bigot, L. (2000). Guide méthodologique pour l'élaboration de cartes de vulnérabilité dês zones côitères de Océan Indien Réalisé par Créocéan, I fremer et Arvan pour le compte de l'UNESCO/IOC et lê PRE-COI/UE. *IOC Manuals and Guides*, No 38.

Fingas, M., & Brown, C., 1997, Remote sensing of oil spill, *Sea Technology*, vol. 38, pp. 37-46.

Fingas, M. & Brown, C. (2000), "Review of Oil Spill Remote Sensing", In *Proceedings of the Fifth International Conference on Remote Sensing for Marine and Coastal Environments*, Environmental Research Institute of Michigan, Ann Arbor, Michigan, pp. I211-218.

Fingas, M. & Brown, C. (2005). An update on oil spill remote sensors. In: *Proc. 28th Arctic and Marine Oil Spill Program (AMOP) Tech. Seminar*, Calgary, Canada, pp. 825-860.

Gonçalves F.; Souza Filho, P.; Paradella, W. & Miranda, F. (2009). Fusão de dados multisensor para a identificação e o mapeamento de ambientes flúvio-estuarinos da Amazônia. *Revista Brasileira de Geofísica*, vol. 27, pp. 57-67.

Gundlach, E. & Hayes, M. (1978). Classification of coastal environments in terms of potential vulnerability to oil spill impact. *Marine Technology Society Journal*, vol. 12, pp. 18-27.

Green, E. 2000. Satellite and airborne sensors useful in coastal applications. In: Edwards A.J. (Ed.) *Remote Sensing Handbook for Tropical Coastal Management*, Costal Management Sourcebooks 3, France: UNESCO Publishing, 41-56pp. ISBN: 9231037366

Grüner, K.; Reuter, R. & Smid, H. (1991). A new sensor system for airborne measurements of maritime pollution and of hydrographic parameters. *Geojournal*, vol. 24, n° 1, pp. 103-117.

Herz, R. (1991). *Manguezais do Brasil*. IOUSP, São Paulo, São Paulo. p.227.

IBAMA. 2011. Informações vetorial sobre porto no Brasil. Avaiable in: siscom.ibama.gov.br

IBGE. 2011. *Sinopse do censo demográfico*. Ministério do Planejamento, Orçamento e Gestão, Rio de Janeiro, pp. 261.

Jha, M.N.; Levy, J. & Gao, Y. (2008). Advances in remote sensing for oil spill disaster management: state-of-the-art sensors technology for oil spill surveillance, Sensors, vol. 8, pp. 236-255.

Jensen, J. (1996). *Introductory digital image processing*, 2 ed., Prentice Hall, London, 318p. ISBN: 0131453610.

Kates, R., Hohenemser, C. & Kasperson, J. (1985) *Perilous Progress: Managing the Hazards of Technology*, Westview Press, Boulder, CO, ISBN: 0813370256

Lima, M.; Dias-Brito, D. & Milanelli, J. (2008). Environmental sensitivity mapping for oil spills in Ilhabela, São Paulo, *Revista Brasileira de Cartografia*, N° 60, pp. 145-154.

Malthus, T. & Mumby, P. (2003). Remote sensing of the coastal zone: an overview and priorities for future research, *International Journal of Remote Sensing*, vol. 24, N°13, pp. 2805–2815.

Mastin, G.; Mason, J.;. Bradley, J.; Axline, R. & Hover, G.(1994). Comparative Evaluation of SAR and SLAR, In: *Proceedings of the Second Thematic Conference on Remote Sensing for Marine and Coastal Environments: Needs, Solutions and Applications*, ERIM Conferences, Ann Arbor, Michigan, pp. I-7-17.

MEA, Millenium Ecossystem Assessment. (2005). Ecossystem and human well-being. Island Press, Washington, DC, ISBN 1-59726-040-1.

Meade, R.; Dunne, T.; Richey, J.; Santos, U. & Salati, E. (1985). Storage and remobilization of suspended sediment in the lower Amazon River of Brazil. *Science*, 228, pp. 488-490.

MMA (2002) Especificações e normas técnicas para elaboração de cartas de sensibilidade ambiental para derramamentos de óleo. MMA, Brasília.

Muehe, D. & Neves, C. (1995). The implications of sea level rise in the Brazilian coast: A preliminary assessment. *Journal of Coastal Research*, vol. 14, pp.54-78.

NOAA. (2002). National Oceanic and Atmospheric Administration. *Environmental Sensitivity Index Guidelines*. Office of Response and Restoration Technical Memorandum No 11, Seattle, USA.

Novaes, R.; Tarouco, J.; Rangel, M. & Dias, L.(2007). *XIV Simpósio Brasileiro de Sensoriamento Remoto*, Florianópolis, Brasil, pp. 4089-4096.

Noernberg, M. & Lana, P. (2002). A sensibilidade de manguezais e marismas a impacto por óleo: fato ou mito? Uma ferramenta para avaliação da vulnerabilidade de sistemas costeiros a derrames de óleo. *Geografares*, vol.3, pp. 109-122.

Pernetta, J. & Elder, D. (1992). Climate, sea level rise and the coastal zone: Management and planning for global changes. *Ocean & Coastal Management*, vol. 18, N°1, pp. 113-160.

PETROBRAS. (2002). *Manual Básico para Elaboração de Mapas de Sensibilidade Ambiental a Derrames de Óleo no Sistema Petrobras: Ambientes Costeiros e estuarinos*. PETROBRAS, Rio de Janeiro.

Poletto, C. & Batista, G. (2008). Sensibilidade ambiental das ilhas costeiras de Ubatuba, SP, Brasil. *Revista Ambiente & Água – An interdisciplinary Journal of Applied Science*, vol. 3, N° 2, pp. 106-121.

Renn, O. (2008). *Risk governance: coping with uncertainty in a complex world*. 1°ed, Earthscan, ISBN 978-1-84407-292-7, USA.

Rocha-Oliveira, T.; Klein, A.; Petermann, R.; Menezes, J. & Sperb, R. (2008). Determinação do índice de sensibilidade do litoral (ISL) ao derramamento de óleo, para região sudeste e sul do estado de Santa Catarina. *Brazilian Journal for Aquatic Sciences and Technology*, vol. 12, N°2, pp. 91-114.

Rodrigues, S. & Souza Filho, P. (2011). Use of Multi-Sensor Data to Identify and Map Tropical Coastal Wetlands in the Amazon of Northern Brazil, *Wetlands*, vol. 31, 11-23pp.

Rodrigues, J. & Szlafsztein, C. (2010). Análise de vulnerabilidade do núcleo urbano da Vila do Conde frente a vazamento de óleo. *Olam: Ciência & Tecnologia*, vol 10, pp. 125-142.

Romero, A.; Riedel, P. & Milanelli, J. (2010). Environmental Oil Spills Sensitivity Mapping of the Cananéia-Iguape Estuarine System, Southern Coast of São Paulo, *Revista Brasileira de Cartografia*, N° 62, special edition 01, 299-238.

Salisbury, J.; D'aria, D. & Sabins, F. (1993). Thermal Infrared Remote Sensing of Crude Oil Slicks. *Remote Sensing of Environment*, vol. 45, pp. 225-231.

Santos, V.; Polidori, L.; Silveira, O. & Figueiredo Jr., A. (2009). Aplicação de dados multisensor (SAR e ETM+) no reconhecimento de padrões de uso e ocupação do solo em costas tropicais – costa Amazônica, Amapá, Brasil. *Brazilian Journal of Geophysics*, vol. 27, supl. 1, pp. 39-55.

Silva, D.; Amaro, V.; Souto, M.; Nascimento, M. & Pereira, B. (2010). Geomorfology of a High Sensitive Area on Potiguar Basin (NE Brazil). *Journal of Integrated Coastal Zone Management*, vol. 10, N°4, pp. 545-566.

Silva, A.; Klein, A.; Petermann, R.; Menezes, J.; Sperb, R. & Gherardi, D. (2008). Índice de sensibilidade do litoral (ISL) ao derramamento de óleo, para a ilha de Santa

Catarina e áreas do entorno. *Brazilian Journal for Aquatic Sciences and Technology,* vol. 12, N°2, pp. 73-89.

Smith, K. & Petley, D. (2008). *Environmental Hazards Assessing risk and Reducing disaster.* Routledge: Londres. ISBN: 0-203-88480-9.

Souto, M.; Castro, A.; Grigio, A.; Amaro; V. & Vital, H. (2006). Multitemporal analysis of geoenvironmental elements of the coastal dynamics of the region of the Ponta do Tubarão, city of Macau/RN, on the basis of remote sensing products and integration in GIS. *Journal of Coastal Research,* vol. 39, pp. 1618-1621.

Souza, C.; Castro, A. ; Amaro, V. & Vital, H. (2005). Sistema de Informações Geográficas para o monitoramento de derrames de óleo no litoral norte do estado do Rio Grande do Norte. In: *XII Simpósio Brasileiro de Sensoriamento Remoto,* Goiânia, Brasil, vol. 1, pp. 2383-2388.

Souza Filho, P. & Elrobrini, M. (2000). M. Coastal Zone Geomorphology of the Bragança Area, Northeast of Amazon Region, Brazil. *Revista Brasileira de Geociências,* vol. 30, No 3, pp. 518-522.

Souza Filho, P. & Paradella, W. (2002). Recognition of main geobotanical features in the in the Bragança mangrove coast (Brazilian Amazon Region) from Thematic Mapper and RADARSAT-1 Data. *Wetlands Ecology and Management,* vol 10, pp.123-132.

Souza Filho, P.; Miranda, F.; Beisl, C.; Almeida, E. & Gonçalves, F. (2004). Environmental sensitivity mapping for oil spill in the Amazon coast using remote sensing and GIS technology. In: *International Geoscience and Remote Sensing Symposium,* IEEE Geocience and Remote Sensing, Ancorage,Alaska.

Souza Filho P. & Paradella, W. (2005). Use of RADARSAT-1 fine mode and Landsat 5 TM selective principal component analysis for geomorphological mapping in a macrotidal mangrove coast in the Amazon Region. *Canadian Journal Remote Sensing,* vol 31, pp. 214–224.

Souza Filho, P. (2005). Costa de Manguezais de Macromaré da Amazônia: Cenários Morfológicos, Mapeamento e Quantificação a partir de Dados de Sensores Remotos. *Revista Brasileira de Geofísica,* Rio de Janeiro, vol. 23, No 4, pp. 427-435.

Souza Filho P.; Gonçalves F.D.; Beisl C.; Miranda F.; Almeida E. & Cunha E. (2005a). Sistema de Observação Costeira e o Papel dos Sensores Remotos no Monitoramento da Costa Norte Brasileira, Amazônia. *Revista Brasileira de Cartografia,* 57, pp. 79 - 86.

Souza Filho, P.; Paradella, W. & Silveira, O. (2005b). Synthetic Aperture Radar for Recognition of Coastal Features in the Wet Tropics: Applications in the Brazilian Amazon Coast. *Boletim do Museu Paraense Emilio Goeldi.* Zoologia, Belém, vol. 1, No 1, pp. 149-154.

Souza Filho, P.; Prost, M.; Miranda, F.; Sales, M.; Borges, H.; Costa, F.; Almeida, E.; & Nascimento Jr.; W.(2009a). Environmental sensitivity index (esi) mapping of oil spill in the amazon coastal zone: the piatam mar project. *Brazilian Journal of Geophysics,* vol. 27, supl. 1, pp. 7-22.

Souza Filho P.; Gonçalves, S.; Rodrigues, F. Costa, R. & Miranda, F. (2009b). Multi-sensor data fusion for geomorphological and environmental sensitivity index mapping in the Amazonian mangrove coast, Brazil. *Journal of Coastal Research,* vol. 56, pp. 1592-1596.

Silva, C.; Souza Filho, P. & Rodrigues, S. (2009). Morphology and modern sedimentary deposits of the macrotidal Marapanim Estuary (Amazon, Brazil). *Continental Shelf Research*, vol. 29, pp. 619-631.

Szlafsztein, C. (2000). Coastal management on the state of Pará and the MADAM project research. In: *German-Brazilian Workshop on Neotropical Ecosystems – Achievements and Prospects of Cooperative Research*, Hamburg, Germany.

Szlafsztein, C. & Sterr H. (2007). A GIS-based vulnerability assessment of coastal natural hazards, state of Pará, Brazil. *Journal of Coast Conservation*. vol. 11, pp. 53-66.

Teixeira, S. & Souza Filho, P. (2009). Mapeamento de ambientes costeiros tropicais (Golfão Maranhense, Brasil) utilizando imagens de sensores remotos orbitais. *Revista Brasileira de Geofísica*, vol. 27, pp. 69-82.

Turner, B., Clark, W., Kates, R., Richards, J., Mathews, J. & Meyer, W. (1990) The Earth as Transformed by Human Action, Cambridge University Press, Cambridge, MA. ISBN: 978521446303

Viles, H. & Spencer, T. (1995). *Coastal Problems Geomorphology, Ecology and Society at the Coast*. Edward Arnold, London. ISBN: 0340625406

Wieczorek, A.; Dias-Brito, D. & Milanelli, J. (2007). Mapping oil spill environmental sensitivity in Cardos Island State Park and surroundings areas, São Paulo, Brazil. *Ocean & Coastal Management*. Vol.50, 872-886 pp.

Wisner, B.; Blaikie, P.; Cannon, T. & Davis, I. (2004). *At risk: natural hazards, people´s vulnerability and disasters*, 2° ed., Routledge, London, ISBN 0415252164.

# 5

# Predictability of Water Sources Using Snow Maps Extracted from the Modis Imagery in Central Alborz, Iran

Seyed Kazem Alavipanah[*], Somayeh Talebi and Farshad Amiraslani

*Faculty of Geography, Department of Cartography, University of Tehran, Tehran*
*Iran*

## 1. Introduction

### 1.1 Snow reserves and remote sensing

Snow reserves in mountainous basins are important and reliable water resources in Iran. Identification of their quality is necessary because of an increasing value of freshwater and utilization of water recourses. About 60 percent of surface water and 57 percent of ground water sources in Iran flows in snowy regions (Rayegani, 2005). The water produced from snowmelt process provides soil water, ground water reserves and water in lakes and rivers. Since snow cover is one of the most important sources of provided water, an accurate prediction and timing of snow runoff is necessary for the efficient management and decision- making in water supply.

The science of snow hydrology, compared to other branches of hydrology science, has a relatively shorter history due to difficulties accompanied with snow measurement. The correct analysis of snow issues needs a set of observations and statistics in snow-gauging. Currently, however, there are no regular and comprehensive snow measurement procedures in most parts of Iran. Measurements are only limited to those snowy basins recharging important dams; even these measurements are carried out in scattered points rather than an entire dam catchment area.

The measurement range of these stations is limited to 2000-3000 m asl heights. Thus, in mountainous Iran, current distribution of stations would not seem to be adequate. In such conditions, study of snow reserves and identification of snow melting trend in most basins would be accompanied with limitations. Consequently, measuring snow cover using ground methods will be difficult and costly. Remote sensing technology has many applications in various environmental and earth resources studies including ice and snow research. These applications have been increased recently as a result of unique technical advantages such as multi-temporal imagery acquired in various wavelengths, extent of spatial coverage, and improvement of computer hardwares for interpretation and extraction of information. Regarding snow research, remote sensing technology can provide

[*] Corresponding Author

continuous information layers with higher accuracy and lower cost compared to the ground survey, so it can fill the information gaps in snow hydrological statistics. However, using ground data can increase the efficiency of remotely-sensed measurement of snow-gauging. Satellites are appropriate tools for gauging snow coverage, because of high reflection of snow that creates proper contrast to most of natural surfaces (with the exception of clouds). Therefore, using satellite imagery and GIS modeling one can produce snow-cover maps, assess the changes in snow cover area with various time series, discriminate snow from other features, and model it in a catchment area. These simplify decision-making process for engineers and hydrology managers.

One of the important issues in remotely-sensed snow-gauging is the selection of sensor. Some of optical sensors that have ever used in snow-measuring include sensors mounted to satellites namely TIROS-1 (1960), ESSA_3, NOAA (1996), LANDSAT (MSS and ETM), and MODIS (2000). Since each sensor has unique properties, a sensor with appropriate spectral, temporal and spatial resolution for snow-gauging must be selected. Since snow is a phenomenon with noticeable surface changes over time, it is necessary to select a sensor that produces proper multi-temporal series. Snow-gauging is done in vast areas, and snow surface is generally even; therefore, MODIS is an appropriate imagery for this purpose. From the view point of spectral resolution, MODIS is one of the best optical sensors for studying snow and discrimination of snow from phenomena such as cloud which has similar spectral reflectance.

One of the purposes of designing of MODIS is a global identification of various types of clouds; hence, several bands have been considered for it to identify various types of cloud cover, optical thickness, effective radius and thermal phase (King et al., 2004).

NASA (National Aeronautics and space administration) launched TERRA satellite to space on December 18th 1999, and MODIS as one of the five sensors mounted on TERRA transferred the first information to Earth on February 24th 2000. MODIS has 36 various bands in visible, infra-red and thermal parts of electromagnetic spectrum including 2 visible bands with 250 m resolution, 5 infra-red bands with 500 m resolution, and 29 thermal bands with 1000 m resolution (Hall et al., 2000).

## 1.2 A review on remotely-sensed snow measurement

Various methods have been used to estimate snow surface such as classification methods, threshold limit, decision-based methods, etc. One of the most applicable algorithms used to estimate snow surface is MODIS snow map algorithm. It was introduced in 1998 as a decision-based algorithm which uses group tests of threshold limit for detection of snow. This algorithm has very small volume from the calculation viewpoint and simple from the conceptual viewpoint, thus user can track how product has been created. In addition, this algorithm has an appropriate efficiency with global application (Hall et al., 1998).

Totally the properties of this algorithm include:

1.  The precision of this method for various types of snow-covered surfaces for identification of snow surface is higher than other methods such as supervised classification, unsupervised classification and sub-pixel methods provided that atmospheric correction is considered (Dadashi Khaneghah, 2008).

2.    This method is a completely automatic algorithm.
3.    This algorithm is applicable for all regions in the globe.
4.    This method is simple, accurate and easy to understand.

Snow Map algorithm uses normalized subtractive index (NDSI). Lee et al. (2001) compared MODIS snow maps created with NDSI index with maps prepared by National Operational Hydrologic Remote Sensing Center (NOHRSC, prepared automatically by GOES and NOAA images) in upper region of Rio Grande reservoir. In NOHRSC, the teta algorithm is used. In this algorithm, two classified images are subtracted to identify snow surface. In teta algorithm, two separate threshold limit is introduced for each image. Lee et al. (2001) concluded that both images are affected by cloudy condition and the main error is cloud coverage. They also mentioned that maps produced from MODIS were more accurate than the above-mentioned maps.

Ault et al. (2006) concluded that MOD10-L2 snow surface product, MODIS sensor, in clear sky condition had the highest accuracy. They showed that the highest error was associated with those conditions that snow depth was lower than 1 cm; thus the higher was the snow depth, the higher was the accuracy. Hall et al. (2000) also showed a low accuracy of low-mass and patchy snows in New England. Klein and Barnett (2003) carried out a snow cover study using MODIS in Rio Grande reservoir during the 2000-2001 period and compared their results to the ground-measuring methods such as snowpack telemetry (SNOTEL) and NOHRSC models. They ultimately concluded that the highest error associated with maps prepared by MODIS was related to the beginning and end of snowfall period. They showed that when the surface was completely covered by snow with no mixed cloud, ground survey or SNOTEL had the highest accuracy.

It can be mentioned that MODIS sensor and NDSI index are appropriate in snow map preparation, although cloud coverage and classification are regarded as constraints (Klein and Barnett, 2003; Zhou et al., 2005). In fact, in spite of various advantages, Snow Map algorithm has some limitations due to inseparability of snow cover from cloud and similarity of cloud behaviours to snow cover. This algorithm cannot completely distinguish clouds from snow (of course, this problem is relatively removable by using Cloud Mask algorithm). Also, this algorithm cannot detect coastal terrains which are similar to snow from viewpoint of whiteness and brightness. However, temperature can act as factor to discriminate snow from these terrains using MODIS bands 31 and 32. Since Cloud Mask and thermal Mask are used before applying algorithm, some error sources in snow map algorithm will be removed (Taghvakish, 2005; Adhami, 2005).

In snow-gauging using satellite imagery, the existence of cloud is problematic due to the following reasons (Riggs and Hall, 2002): first, clouds conceal Earth information; second, clouds create shades on area and change reflectance. Indeed, if clouds cannot be detected well, they will reduce the accuracy of snow map.

Clouds and snow have generally similar spectral reflectional properties in range of visible and infra-red spectra, so thermal properties is not enough for distinguishing them from each other as clouds may be cooler or hotter than snow (Singh and Singh, 2001). In order to detect clouds, a procedure called Cloud Mask algorithm is being used. Akerman et al. (1998) introduced MOD35 Cloud Mask algorithm. MOD35 algorithm is based on obstruction of Earth surface affected by cloud or dust particles that identifies water body, land and atmosphere (Strabala,

2003). In this process, based on land type, geographical position and available data, Cloud Mask algorithm uses 14 bands amongst 36 bands of MODIS to test 18 spectral and spatial features (Hall and Riggs, 2002). However, this procedure was modified by Hall and Riggs (2002) who presented a new version of Cloud Mask algorithm (Liberal). This algorithm can analyse the pixels located under thin and transparent clouds (Zhou et al., 2005; Ault et. al., 2006). This procedure identifies the darkness and if it faces to such darkness, it means that sun angle is higher than 85°. This algorithm is called Liberal Cloud Mask algorithm. In fact, Liberal Cloud Mask algorithm functions as subset of spectral tests of old Cloud Mask algorithm (MOD35) and uses 7 bands of MODIS and set 4 criteria (Hall and Riggs, 2002, 2004).

## 2. Material and methods

### 2.1 Study area

The Central Alborz mountainous range extends from 49° 5´- 53° 5´ longitude to 35° 5´- 37° 2´ latitude. Its area is about 40,000 km² and covers 64 sub-basins. The lowest, highest and average altitude of the basin is 48 m, 5671 m and 1870 m, respectively. The minimum, maximum and average slope is 5%, 25.4% and 23.56%, respectively. The main slope aspect of this region is directing towards north and south. Climatically, the region is classified to three classes as temperate in north, cold in center, and semi-arid in south. Geological structures mostly consist of mild outcrops that are inconsistent with general trend of east-west. In western part of Alborz, the structures have northwest-southeast trend, but in eastern part, the structures have northeast-southwest trend. These two inconsistent trends cross each other in Central Alborz.

### 2.2 Data

### 2.2.1 MODIS Data, TERRA satellite

MODIS encompasses noticeable number of spectral and thermal bands with narrow width, high radiometric resolution, proper width and collecting time, powerful and accurate calibration, and diverse land resolution (MODIS Home page) (http://modis.gsfc.nasa.gov).

In many cases, MODIS provides satellite snow-gauging requirements and therefore these data were used. In this research, images were provided from website (https://wist.echo.nasa.gov/wist) according to Table 1.

| Year | 2006 | 2006 | 2006 | 2006 | 2006 | 2006 |
|------|------|------|------|------|------|------|
| Month | February | February | march | march | march | march |
| Day | 21 | 25 | 4 | 8 | 13 | 15 |

Table 1. Temporal table applied images from MODIS sensor data

The imagery used in this research include MOD02 and MOD09. The MOD02 imagery include 36 bands while MOD09 include 6 bands. In MOD09GA imagery, atmospheric corrections have been done based on 6sv model, as one of the best models in atmospheric corrections with minimum error, suitable for measuring snow surface and detecting cloud from snow (Vermote and Kotchenova, 2008). In MOD09 imagery, corrections have been implemented in way that atmospheric diffusion and reflection were minimum. Since data with higher wavelength are being less influenced by aerosols, suspended particles and non-selective diffusion phenomenon, thermal bands of MOD02 imagery were used.

### 2.2.2 Digital Elevation Model

Digital Elevation Model (DEM) obtained from SRTM Shuttle was used. These data that was in format of GeoTiff had Lambert image system. For transformation of this data to UTM coordinates system, PCI Geomatica software was used.

### 2.2.3 Ground stations data

Snow-gauge station data were obtained from Water Organization- Department of Surface Water. Snow-gauge stations of Central Alborz are located in five basins namely Lattian, Lar, Taleghan, Karaj and Golpayegan basins. In some cases, ground data survey time was not consistent with the time of image acquisition. For solving this problem, in those dates that no ground statistics were available, previous and next day's information were interpolated. Of course interpolation was carried out for those stations where sampling time was close to image acquisition date, and the station had snow cover during the period (February and March). Among sampled stations, 18 stations with above-mentioned conditions were selected.

It is necessary to mention that snow depth data has been used to examine the presence or absence of snow cover. Those gauging stations located far from human interfering features (e.g. buildings) were selected and their snow depth measured. The snow surface was defined an area where surface is regular and even, with the minimum wind blowing effect to increase measurement accuracy (Pfister and Schneebli, 1999). The output spatial resolution of snow map algorithm and Cloud Mask Algorithm is one kilometre. Around each station, up to two kilometres was regarded to include 9 to 13 pixels covering snow. The most repeated pixel shows snowy or not snowy condition. Finally, snow surface obtained from Snow Map Algorithm with and without Liberal Cloud Mask was compared to ground data. Figure 1 illustrates distribution pattern of snow-gauge stations within five basins in Central Alborz.

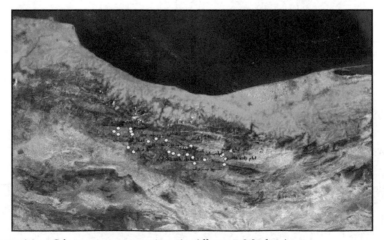

Fig. 1. The position Of snow gauge stations in Alborz-e-Markazi

## 2.3 Research method

Figure 2 illustrates overall flowchart of this research methodology.

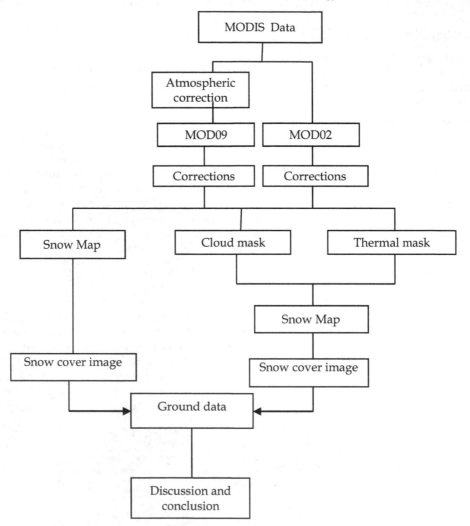

Fig. 2. The flowchart of this research methodology

### 2.3.1 Snow map algorithm

Snow map algorithm benefits from Normalized Difference Snow Index (NDSI). Because of low reflection of snow in infrared bands and high reflection in visible bands, NDSI can be useful for discrimination of snow from other phenomena. NDSI is calculated by equation below (Hall et al., 1998):

NDSI = (band4-band6) / (band4+ band6)

Snow map algorithm includes following thresholds:

If NDSI $\geq$ 0.4; and MODIS band2 > 11%; and MODIS band 4 $\geq$ 0.4.

NDSI index is used to recognize snow and ice and also to differentiate between cumulus and ice or snow. In fact this index represents relative differential reflectance value of visible and short wave-infrared channels emitted from snow. Pure snow has a high NDSI value but other materials such as soil, smoke, etc cause a NDSI reduction. The mentioned threshold for band 4 of MODIS is a key tool to prevent identifying pixels with low reflection, for example dark Cypress, instead of snow. Water and cloud are separable using mentioned threshold for band 2 of MODIS and finally, NDSI has the key role in investigating snow (Hall et al., 1998).

Necessary eligibilities of pixels for applying snow map algorithm are as follows (Riggs et al., 2003):

- Pixels should have level1B reflection (geo-reference process and radiometric correction should be done),
- They should belong to terrestrial region or water bodies surrounded by lands,
- Imagery should be taken in day light,
- Imagery should not be covered by cloud (applying cloud mask),
- Their approximate temperature should be less than $283^0$ K (applying temperature mask)

### 2.3.2 Cloud isolation

Liberal cloud masking just uses 7 out of 36 MODIS bands as well as 4 out of 18 old cloud masking algorithm criteria. Before performing snow map as one of the preprocessing steps, liberal cloud masking was applied. With regards to spectral resemblances between snow and cloud, applying mask on image is inevitable. In liberal cloud masking, a pixel will be considered as cloud provided that it covers one of the following criteria:

1. High cloud index introduces it as cloud
2. Heat difference index consider it as cloud
3. Visual bands reflection index proves the existence of cloud when reflection of band (1/625, 1/628, μm) is more than 20 percent and visual band threshold is applied.
4. NDSI $\geq$ 0.4 and reflection of band 6 is more than 20 percent (Riggs and Hall, 2002).

In this research, since the study area is located on terrestrial area and consequently discrimination of snow from cloud is very important and also all MODIS imagery were taken in day time, thresholds related to terrestrial region in day time were applied. Furthermore, water bodies were eliminated before image processing.

### 2.3.3 Heat masking

Heat masking is the final step before using snow map algorithm. This method was introduced on 3rd October 2001 and resulted in eliminating many of incorrect land cover classified as snow. In MODIS version 3, a threshold of $277^0$ K was used whereas in version 4 this value increased to $283^0$ K. Every individual pixel of band 31 with a temperature more than threshold of version 4 is not classified as snow (Kamanpoon, 2004). Heat masking is used to remove ambiguity between snow and other phenomena such as water bodies, sand

and cloud (Zhou et al., 2005). In this part using calculated apparent temperature for band 31 and applying 283⁰K threshold, heat masking is performed after new cloud masking algorithm and before snow map algorithm.

## 3. Results

Images resulted from snow map algorithm before and after applying Liberal cloud mask related to February and March are illustrated in Figure 3. Right column shows images before liberal cloud masking and middle column show them after masking. Left column shows false color images which are made by combining visual and infrared channels of MODIS according to method introduced by Miller et al. (2004). Lands without snow cover, with snow cover, low height clouds and higher clouds appear as green, white, yellow and violet tones, respectively. False color image help to recognize cloudy regions on image as well as cloud height.

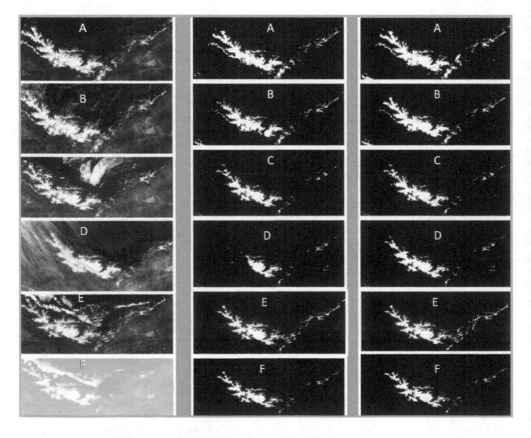

Fig. 3. Snow area before applying the liberal cloud mask (right column) and after applying the Liberal cloud mask (middle column) and false color composite (left column).

## 4. Discussion

In this part, using NDSI, topographic data and data gathered from snow measurement stations, snow map algorithm alone and together with Liberal cloud masking were separately interpreted. Ground-based snow measurement data and their corresponding points on images resulted from snow map algorithm as well as images resulted from snow map algorithm together with Liberal cloud masking in different dates are illustrated in table 2. In fact error matrix is drawn for each image and results have been surveyed.

| Date of acquisition | Snow map | Snow | No snow | Total | Accuracy % | Snow map (masking) | Snow | No snow | Total | Accuracy % |
|---|---|---|---|---|---|---|---|---|---|---|
| 13/3/2006 | Snow | 12 | 2 | 14 | 77 | Snow | 12 | 0 | 12 | 100 |
| | No snow | 0 | 4 | 4 | 100 | No snow | 0 | 6 | 6 | 100 |
| | Total | 12 | 6 | 18 | | Total | 12 | 6 | 18 | |
| | Accuracy % | 100 | 66 | | 88 | Accuracy % | 100 | 100 | | 100 |
| 15/3/2006 | Snow | 12 | 2 | 14 | 85 | Snow | 12 | 0 | 12 | 100 |
| | No snow | 0 | 4 | 4 | 100 | No snow | 0 | 6 | 6 | 100 |
| | Total | 12 | 6 | 18 | | Total | 12 | 6 | 18 | |
| | Accuracy % | 100 | 66 | | 88 | Accuracy % | 100 | 100 | | 100 |

Table 2. Evaluation of accuracy of snow gauge obtained from snow map algorithm and snow map algorithm with attending the Liberal cloud mask using earth data in 6th of February, 13th and 15th of March

Results demonstrate that in both images of 13th and 15th March in which snow map was applied, the number of points classified as snow is more than the time when applying snow map algorithm; adding cloud masking to snow map algorithm reduces this number. It means that regions which are incorrectly classified as snow by snow map algorithm can be categorized as cloud after adding cloud masking. Furthermore, no snow regions identified as snow in snow-gauging station and snow map algorithm with Liberal cloud masking are more than those no snow regions that are not classified as snow without applying Liberal cloud masking. So it can be concluded that snow map algorithm shows some regions as snow despite the fact that they are clouds. However, cloud masking can detect them and classify as cloud. Error matrix demonstrates that accuracy of snow map algorithm increases by applying cloud mask (Riggs and Hall, 2002; Ault et al., 2006; Hall and Riggs, 2007). Overall, results from snow map algorithm together with Liberal cloud masking are more compatible with data gathered from ground-based stations.

One of the factors affecting accuracy of snow detection is clouds which cover snow surface. These clouds are distinguishable by Liberal cloud masking provided that they are transparent and thin (Riggs and Hall, 2002; Ault et al., 2006). In images related to 21st February and 8th March, the observed cloud is thick and far from the Earth. False color images show that clouds are far from the Earth surface in both mentioned images so they can be detected and classified correctly by Liberal cloud masking. However, there is snow under these clouds and should be considered as snow. Data from ground-based snow measurement and their corresponding points on images related to 21st February and 8th March resulted from snow map algorithm before and after applying Liberal cloud masking is shown in Table 3. As is shown in Table 3, field survey data are different from results obtained as a result of snow map algorithm together with Liberal cloud masking. In this situation, considering neighborhood effect, topographic factors and false color images, clouds over snow can be distinguishable and classify them as snow. Of course, neighborhood and topographic factors can be helpful when the cloud is smaller that total area of snow.

| Date of acquisition | Snow map | Snow | No snow | Total | Accuracy % | Snow map (masking) | Snow | No snow | Total | Accuracy % |
|---|---|---|---|---|---|---|---|---|---|---|
| 21/2/2006 | Snow | 15 | 2 | 17 | 77 | Snow | 12 | 0 | 12 | 100 |
| | No snow | 0 | 1 | 1 | 100 | No snow | 3 | 3 | 6 | 50 |
| | Total | 15 | 3 | 18 | | Total | 15 | 3 | 18 | |
| | Accuracy % | 100 | 66 | | 88 | Accuracy % | 83 | 16 | | 83 |
| 8/3/2006 | Snow | 12 | 2 | 14 | 85 | Snow | 8 | 0 | 8 | 100 |
| | No snow | 0 | 4 | 4 | 100 | No snow | 4 | 6 | 10 | 55 |
| | Total | 12 | 6 | 18 | | Total | 1 | 6 | 18 | |
| | Accuracy % | 100 | 66 | | 88 | Accuracy % | 66 | 100 | | 77 |

Table 3. Evaluation of accuracy of snow gauge obtained from snow map algorithm

There is a negligible difference before and after applying Liberal masking images covered by low height clouds (e.g. image of 4th March) (Table 4). It means that in this situation snow map algorithm with and without liberal cloud masking has the same result. So it can be concluded that snow map algorithm is able to detect low height clouds because the spectral diagram of low height clouds are different from that of snow in visual and infrared spectrum range. Data gathered from ground-based snow measurement and its corresponding points on images resulted from snow map algorithm as well as images resulted from snow map algorithm together with liberal cloud masking is shown in Table 4.

| Date of acquisition | Snow map | Snow | No snow | Total | Accuracy % | Snow map (masking) | Snow | No snow | Total | Accuracy % |
|---|---|---|---|---|---|---|---|---|---|---|
| 4/3/2006 | Snow | 11 | 0 | 11 | 100 | Snow | 11 | 0 | 11 | 100 |
| | No snow | 1 | 6 | 7 | 85 | No snow | 1 | 6 | 7 | 85 |
| | Total | 12 | 6 | 18 | | Total | 12 | 6 | 18 | |
| | Accuracy % | 91 | 100 | | 94 | Accuracy % | 91 | 100 | | 94 |

Table 4. Evaluation of accuracy of snow gauge obtained from snow map algorithm and snow map algorithm with attending the Liberal cloud mask using earth data in 4th of March

In order to show NDSI ability in isolation of cloud from snow, variation range of NDSI in regions which are identified as cloud using new cloud masking is compared with variation in regions where classified as snow by use of snow map algorithm. As it can be found in diagram NDSI variation in regions where identified as cloud and regions where classified as snow has some overlaps so NDSI cannot distinguish between snow and cloud in these regions (Figure 4).

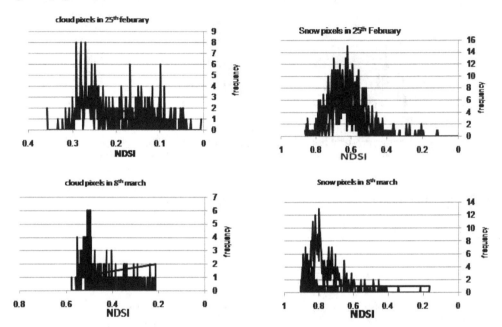

Fig. 4. The variation range of DSI in cloudy and snowy area on images of 25th February and 8th March

Figure 5 shows comparison of NDSI variation for image taken in 4th March which has only low height clouds in regions which are identified as cloud using new cloud masking with variation in regions where classified as snow using snow map algorithm. As it can be seen in diagram NDSI variation in regions where identified as cloud and regions where classified as snow are absolutely separable so NDSI can distinguish between snow and cloud in these regions and act similar to new cloud mask (Figure 5).

As a general rule, the amount of snow increases in higher elevations so if classification of snow and cloud is done perfectly, percentage of pixels related to snow should increase in higher elevation. This rule can be used to evaluate the accuracy of outputs resulted from snow map algorithm alone and together with Liberal mask algorithm. Figure 6 shows the relative frequency of snow pixels in each altitudinal zone. As it can be seen in this Figure, ascending trend occur whenever new cloud mask is applied together with snow map algorithm. In fact, cloud masking leads to the better identification of cloud pixels and prevents these pixels to be classified as snow. However ascending trend will not happen in mentioned diagram when snow map algorithm is used alone because some cloud pixels are categorized as snow incorrectly.

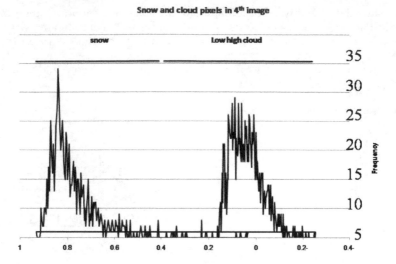

Fig. 5. The variation range of NDSI in cloudy and snowy area on images of 4th March

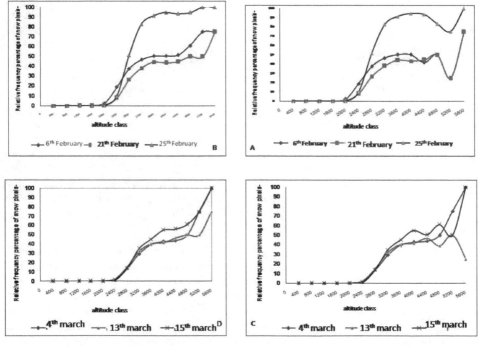

Fig. 6. Relative frequency percentage of snow pixels in each altitude class; right: the obtained images of snow map algorithm; left: the obtained images of snow map algorithm accompanying the Liberal clod mask

## 5. Conclusion

Reviewing data resulted from ground-based snow measurements in addition to results from snow map algorithm and Liberal cloud mask, it can be concluded that snow map algorithm cannot detect some types of cloud and classify them as snow (Zhou et al., 2005; Riggs and Hall, 2002; Ault et al., 2006; Hall and Riggs, 2007), reducing the accuracy of maps produced for snow detection. Clouds which are not detected by snow map algorithm are those include ice particles in high elevations (Taghvakish, 2005). Using Liberal cloud masking can largely solve this problem and prevent some types of clouds to be categorized as snow. The accuracy of maps is increased approximately 10% in comparison with other methods. In images including only low elevation clouds, cloud masking cannot make better results; therefore it can be concluded that these kinds of clouds can be detected by snow map algorithm alone. Also, results from applying NDSI shows that some types of clouds are categorized in the same class as snow, so NDSI cannot distinguish between snow and cloud. However, those clouds in low elevation can be detected from snow.

Altitudinal parameter is another tool in order to evaluate the accuracy of snow map algorithm and Liberal cloud masking. An ascending trend in frequency of snow pixels is evident whenever cloud masking is used in addition to the snow map algorithm.

In summary, it can be said that although low height clouds are separable by snow map algorithm, some types of clouds cannot be detected by snow map algorithm alone and thus, application of cloud masking is inevitable. These are clouds which are in high elevation and include ice particles (Taghvakish, 2005). Finally, in some cases even cloud masking cannot distinguish between snow and ice particles (Ault et al., 2006, Riggs and Hall, 2002 , Taghvakish, 2005).

## 6. References

Ackerman S A, Strabala K I, Menzel P W P, Frey R A, Moeller C C  and Gumley L E, 1998: Discriminating clear sky from clouds with MODIS, *Journal of Geophysical Research*, 103(D24):32,141-32,157.

Adhami S, 2005. Application of remote sensing and geographic information system in snow cover (Agichay). Unpublished MSc Thesis, University of Tabriz (In Persian).

Ault T W, Czajkowski K P, Benko T, et al., 2006. Validation of the MODIS snow product and cloud mask using student and NWS cooperative station observations in the Lower Great Lakes Region. *Remote sensing of Environment* 105: 341-353.

Dadashi Khanegha S, 2008. Appointment of snow cover using image processing techniques. Unpublished MSc Thesis. University of Shahid Beheshti (In Persian).

Hall D K, Tait A B, Riggs G A, Salomonson V V, Chien J, Andrew Y L, and Klein G. 1998. Algorithm Theoretical Basis Document (ATBD) for the MODIS Snow-, Lake Ice- and Sea Ice-Mapping Algorithms, MODIS Algorithm Theoretical Basis Document Number ATBD-MOD-10, NASA Goddard Space Flight Center, 1998.

Hall D K and Riggs G, 2007. Accuracy assessment of the MODIS snow products. *Hydrological Processes* 21:1534-154.

Hall D K, Tait A B, Foster J L, Change A T C, and Allen M, 2000. Intercomparison of satellite-derived snow-cover maps. *Annals of Glaciology* 31:396-376.

Kamanpoon S, 2004. Hydrological modelling using MODIS data for snow covered area in the Northern Boreal Forest of Manitoba, MSc Thesis.

King M D, Platnick S, Yang P, Arnold G T, Gray M A, Riedi J C, Ackerman S A, Liou K, 2004. Remote sensing of liquid water and ice cloud optical thickness and effective radius in the Arctic: Application of airborne multispectral MAS Data. *Journal of Atmospheric and Oceanic Technology* 21: 857-875.

Klein A G and Barnett A C, 2003. Validation of daily MODIS snow cover maps of the Upper Rio Grande River Basin for the 2000-2001 snow year. *Remote Sensing of Environment* 86:162-176.

Lee S, Klein A G and Over T M, 2001. A Comparison of MODIS and NOHRSC snow cover products for simulating streamflow using the Snowmelt RunOff Model in: http://www.modissnowice.gsfc.nasa.gov

Miller S D and Lee T F, 2004. Satellite-Based imagery techniques for daytime cloud/snow delineation from MODIS. *Journal of applied meteorology*. volume 44.

Pfister R and Schneebeli M, 1999. Snow accumulation on boards of different sizes and shapes. *Hydrological Processes* 13:2345-2355.

Hall D k, Foster J L, Robinson D A, Riggs G A, 2004. Merging the MODIS and RUCL monthly snow cover records. In Proceeding of IGARSS 04, Anchorage, AK, September 2004.

Rayegani B, 2005. Investigation on snow cover changes and estimation snowmelt runoff using MODIS data. Unpublished MSc Thesis. Isfahan University of Technology (In Persian).

Riggs G and Hall D K, 2002. Reduction of cloud obscuration in the MODIS snow data product, 59th Eastern Snow Conference, Stowe, Vermonte, USA.

Riggs G A , Haii D K and Salomonson V V, 2003. MODIS snow products users guide. Available at: http://www.modis-snow-ice.gsfc.nasa.gov/sug.pdf.

Strabala K, 1999. MODIS Cloud Mask User's Guide', *User's Guide*.

Singh P and Singh V P, 2001. *Snow and Glacier Hydrology*, Kluwer Academic Publishers, p742.

Taghvakish S, 2005. Supervision of snow cover using satellite images. Unpublished MSc Thesis, University Of Tehran (In Persian).

Vermote E F and Kotchenova S, 2008. Atmospheric correction for the monitoring of land surfaces. . *Journal of Geophysical Research -ATMOSPHERES*, 113(D23), D23S90.

Zhou X, Xie H and Hendrick J M H, 2005. Statistical evaluation of remotely sensed snow cover products with constraints from streamflow and SNOTEL measurements, *Remote Sensing of Environment* 94: 214–231.

# 6

# Remote Sensing Application in the Maritime Search and Rescue

Jing Peng and Chaojian Shi
*Shanghai Maritime University*
*P.R. China*

## 1. Introduction

Maritime search and rescue (MSR- In the maritime publications, the abbreviation for search and rescue is also SAR. Here we use MSR to distinguish it from the abbreviation for Synthetic Aperture Radar.) became an enormous task with the vast growth of marine transportation and other marine activities. In the year of 2006, the MSR centers and maritime authorities in China organized and coordinated 1620 MSR operations, which involved 5322 vessels and 17498 human lives. The past few years have witnessed tremendous changes in the organizations of maritime rescue. A large part of this evolution stems from the involvement on an international scope and the contribution of the advanced technology. However, current maritime search operation, especially searching people over board, depends mostly on human eyes.

SOLAS (International convention for safety of life at sea) convention prescribes that ships must be equipped with GMDSS (Global maritime distress and safety system) equipments, which have improved the search and rescue. However, for many non SOLAS convention ships, such as fishing boats and small crafts, the detection results are not very much satisfied. With the complex sea environment, the searching of distress vessel becomes a nail-biting task. Because of the physiological characteristics of human eyes, it is difficult for the rescuer to find small target in the adverse background lighting, night or dark condition, wave or clustered seas. Continuous long time observation also causes fatigue of human eyes, resulting poor sensitivity of detection. All those factors decay the results of searching operation.

In order to improve the effect of MSR operations during the dark hours or in adverse lighting or sea conditions, remote sensing technique is a potential approach to overcome the limitation of human eyes in MSR, and thereby may hopefully improve the searching performance in complex environment or in a fatigued state of human being. Regarding ship monitoring, compared with shore-base, shipboard or airborne detecting devices, and other visible visible or infrared monitoring methods, the Synthetic Aperture Radar (SAR) remote sensing system possesses the capability of all-time, all weather, extensive and high resolution for detecting ships on the sea. Especially due to its working characteristics of not being limited by the sea surface, weather or human factors, it can detect the sea areas with geographical remote positions and hostile environment which cannot be entered directly.

In this chapter, some remote sensing techniques and algorithms concerned with the MSR are introduced. A Remote Sensing Monitoring System for Maritime Search and Rescue (RS-MSR) is presented. This work is a part of our project—Vision Enhancement System for Maritime Search and Rescue. The main task for the RS-MSR is to acquire general information in a wider scale. The distress ship is detected and located for guiding the search operation. Surrounding ships are also distinguished to coordinate the MSR operation. Some important data such as current and sea state are retrieved to help decision-making of the operation. Section 2 proposes the outline of the remote sensing methods for maritime search and rescue; Guided by the systematic functions and structures of the RS-MSR described in Section 2, Section 3 introduces the related algorithms used in RS-MSR. Section 4 describes the architecture of the remote sensing aided system for maritime search and rescue. The experiment design and the implementation performance are given in Section 5. Finally, Section 6 concludes the paper.

## 2. Outline

The primary role of the remote sensing is to provide a secondary source of information for the MSR operations. A remote sensing monitoring system can, to some extent, overcome the shortcomings and inadequacies of human eyes. It can also improve the searching speed and accuracy, and is of significance in promoting rescue success rate and efficiency. It can aid the rescuers to fulfill the task of search and rescue, especially for small targets, such as persons in distress and life boats, and could provide a good detection and identification result.

### 2.1 Introduction to the system functions

According to the requirement of search and rescue, a Remote Sensing Monitoring System for Maritime Search and Rescue (RS-MSR) is designed. Table 1 illustrates the main functions of a RS-MSR system.

| RS-MSR System demand | Function |
|---|---|
| Satellite transit inquiry | Satellite transit inquiry |
| Ship detection | 1) Ship position detection<br>2) Ship type identification/classification<br>3) Ship size estimation<br>4) Ship heading/direction estimation |
| Sea state analysis | Wave direction and estimation |
| Integrated processing | 1. Non-remote sensing data fusion<br>   - VTS (Vessel Traffic Services) report<br>   - AIS (Automatic Identification System) report<br>2. Performance analysis<br>   - Recognition rate<br>   - Identification/classification rate<br>   - Position error<br>3. Rescue position prediction<br>   - Time prediction<br>   - Ship speed and velocity<br>   - Predict the rescue area according to the heading/direction of the distress ship and the sea state |

Table 1. The main functions of a RS-MSR system required

## 2.2 Algorithm introduction

According to the requirement of the maritime search and rescue, the Synthetic Aperture Radar (SAR) imageries are used for this purpose. The algorithms concerned with the system demand are introduced as follows.

### 2.2.1 Distributed target detection method in the Gaussian scale-space

In the SAR images with high resolution, each target occupies several resolution units to form area target. So detecting the ship target in the SAR images with high resolution should regard the target as distributive target, and the assumption of point target under the traditional radar is not suitable any more. The project here proposes a distributed target detection method in the Gaussian scale-space. The distance relationship among the detected objects is adopted to identify the distributed target. In the situation of hardly estimating the background's scattering distribution or of low SNR (signal-to-noise ratio), this method can realize the distributive target detection more effectively than CFAR method.

### 2.2.2 Ship size category estimation model

In the SAR images with high resolution, the ship target can be divided into three categories according to their dimension, among which small ships($\leq$50m) are represented as point target, while the middle size ship($\leq$100m) as distributive target of single corner and big size ship($\geq$100m) as double corner distributive target. According to the relative positions of the corner and combining the orbit information (resolution ratio and incident angle) of the SAR image, the length, height and the direction of the ship will be worked out.

### 2.2.3 Ship location correction and ship direction estimation method

The imaging geometry of the SAR imagery is slant-range projection. So, due to the geometric distortions, such as layover, foreshortening, and shadows, there exits measurement error between the observed position and its actual location. In this method, using the directional texture of the wake of ship, the convergent point of the wake pattern can be calculated, which is the actual location of the stern. Then the position correcting parameters can be worked out.

A bow wave is the wave that forms at the bow of a ship when it moves through the water. As the bow wave spreads out, it defines the outer limits of a ship's wake. Theoretically, the convergent point of the bow wave' outline must at the extended line of the ship's heading. The vanishing point can be calculated by the Hough transform, and the heading direction of the ship can be calculated according to the coordinates of the bow and the stern.

### 2.2.4 Wave direction estimation based on the partial energy direction

Wave direction estimation can be used to analyze the sea state of the target area and supply basis for search and rescue area decision. This method is set up on the basis of steerable filter, which is a filter set composed of an even-symmetric filter and an odd-symmetric filter. When the orthogonal filter set is rotated to the same orientation of the local texture, the oriented energy reaches its maximum. The orientation corresponding to the maximum oriented energy is defined as the dominant orientation of the local energy at that point. And the main direction of the wave can be estimated.

### 2.2.5 The registration algorithm of SAR image and nautical chart based on Gaussian principle curve

To ensure precise detection and location of the distressed ship in the MSR, the navigational chart and the remote sensing image should be matched beforehand. Because the SAR image and the electronic chart are data from different sensors, the content and intensities of these images are much different from each other. Coastline is a stable and reliable feature for navigation in coast area. However, the deformation between edges extracted from different signals may produce position errors, and the noise in radar signals may greatly influence the edge extraction result. And how to obtain reliable control-points and how to obtain the correct correspondence are the key issues in the registration algorithm. In this chapter, a multi-scaled registration algorithm for SAR image and electronic chart is presented. Based on the scale-space theory, coastlines from the two images are matched in both frequency domain and image domain with continuous scale level.

## 3. Algorithms and methods in the maritime search and rescue

### 3.1 Distributed target detection method in the Gaussian scale-space

In the high resolution SAR images, the large-scale ship and super large-scale ship (>100 meters) are presented as distributed targets. In this algorithm, these targets are detected by the distance relationship between the echo intensity of the masthead light and the ship hull. The detection of distributed target based on location-dependent information can be completed by two-step detection. In the first step, the ship target is characterized in the Gaussian scale-space. This transforms the signal range value into binary. Then, the detection of the singular objects is implemented using constant false alarm rate(CFAR), and record the location of the pixel point whose value is 1. The second step aims at finding the distributed target from the result of the first step detection by applying location-dependent information.

### 3.1.1 Ship characteristic description

Under the ideal condition, in the $z(x)$ of the image which has its background gray value as 0, there is a maxima value $h$ and strong scattered point $f_b(x)$ with width as $w$. Considering the edge effect of the image, we build the mathematical model as follows:

$$f_b(x) = \begin{cases} h(1-(x/w)^2), & |x| \le w \\ 0, & |x| > w \end{cases} \tag{1}$$

The responses of these spot-like targets in Gaussian scale space are presented as $r_b(x,\sigma,w,h)$:

$$\begin{aligned} r_b(x,\sigma,w,h) &= g_\sigma(x) * f_b(x) \\ &= \frac{h}{w^2}[(w^2 - x^2 - \sigma^2)(\phi_\sigma(x+w) - \phi_\sigma(x-w)) \\ &\quad -2\sigma^2 x(g_\sigma(x+w) - g_\sigma(x-w)) - \sigma^4(g'_\sigma(x+w) - g'_\sigma(x-w))] \end{aligned} \tag{2}$$

Among them, $\phi_\sigma(x) = \int\limits_{-\infty}^{x} e^{-\frac{x^2}{2\sigma^2}} dt$ .

In order to departure the ship target from the grey scale space, we define a Gaussian comparison function $e_b(x)$. If the responses of $f(x)$ in Gaussian scale-space is $r(x,\sigma, w, h)$, then we have:

$$e_b(x) = \begin{cases} 1, & r(x,\sigma,w,h) \le f(x), \text{ and } |x| \le w \\ 0, & \text{other} \end{cases} \qquad (3)$$

Then, the detection of the singular target is implemented by way of constant false alarm rate(CFAR).

### 3.1.2 Distributed target detection based on location-dependent

Set the SAR image as a $N_R \times N_A$ matrix, where $N_R$ and $N_A$ represent the dimension of range and azimuth respectively, and use $\{i_{R1}, i_{R2}, \cdots, i_{RK}\}$ as range coordinate and $\{i_{A1}, i_{A2}, \cdots, i_{AK}\}$ as azimuth coordinate separately for convenience.

The detected relative distance among different scattered units can be defined.

$$d(j,k) = \|i_j - j_k\|, k > j, j = 1,2,\cdots, K-1 \qquad (4)$$

where $\|\ \|$ means norm. Equation(4) represents the relative distance between two scattered units $j$ and $k$, and it expresses the location relations among each target pixel point after the Gaussian scale-space detection.

When the range and the azimuth are considered separately, the distance can be defined as

$$d(j,k) = \left( \left| i_{Rk} - i_{Rj} \right|, \left| i_{Ak} - i_{Aj} \right| \right) \qquad (5)$$

Due to the geometric distortions of the SAR image, the ship hull and the mast will occupy several resolution units in the image. Set a candidate ship target $M_0$ ($M_0$ includes the ship hull and the mast), where $M_{0R}$ and $M_{0A}$ represent the size of $M_0$ in the range direction and the azimuth direction respectively. Set a bounding box with size of $M_{0R} \times M_{0A}$ , which is the smallest rectangle containing the detecting ship. Assume that the size of the detecting target is $S_R \times S_A$ and the radar resolution is $\Delta R \times \Delta A$ , then $M_{0R} \times M_{0A} = \dfrac{S_R}{\Delta R} \times \dfrac{S_A}{\Delta A}$ .

Regarding to the distributed targets, at least $M$ scattering points will be detected after the second detection step, and then $M$ can be defined as the distance threshold $d_{th}$ within the distributed target.

$$d_{th} = M = \mu M_{0R} \times M_{0A} \qquad (6)$$

Here $\mu \in (0,1)$ is the confident coefficient and it is determined by the empirical data of the radar echo.

Define the size of the distributed target $T_s$ as:

$$T_s = d(M_0,(0,0)) \tag{7}$$

Equation (7) defines the distance from the position of the bounding box $M_0$ to the origin point (0,0). The distance describes the size of the detecting target, which is related to the form of the distance definition.

According to the size of the bounding box $M_0$ and Equation (6), the number of scattered points (marked as $u$) of each target reference window and their locations (i.e. target location) can be calculated as follow.

$$\left. \begin{array}{ll} u \geq d_{th}, & \text{target} \\ u < d_{th}, & \text{non} - \text{target} \end{array} \right\} \tag{8}$$

Under the definition in Equation (7), $d(T_{position}, i_k)$ and $T_s$ are all two-dimension data, we need define its output. Set $T_s = (T_{sR}, T_{sA})$, $d = (T_{position}, i_k) = (d_R, d_A)$, and then the export definition of $d(T_{position}, i_k) > T_s$ is:

$$d = (T_{position}, i_k) > T_s = \begin{cases} 1, & \text{other} \\ 0, & d_R \leq T_{sR} \text{ and } d_A \leq T_{sA} \end{cases} \tag{9}$$

### 3.1.3 Algorithm realization

According to the above description of the algorithm, to design the integrated detection algorithm based on two-dimension location-dependent information, the procedure is shown as follows:

1. Initially set $k = 1$, $u = 1$, and the position of the target centre is $T_{position} = i_k$;

2. If $d = (T_{position}, i_k) \leq T_s$, the number of pixels contained in the current target is $u = u + 1$, and renew the central position of the current target $T_{position} = \dfrac{(u-1)T_{position} + i_k}{u}$;

3. If $d = (T_{position}, i_k) > T_s$, then create a new target with its central pixel location as $T_{position} = i_k$, $u = 1$;

4. Traversal searches the image.

### 3.2 Ship size category estimation model

In the SAR images with high resolution, the ship target can be divided into two categories according to their dimension. The small ships are represented as single corner target while big ships as double corner distributed targets. According to the relative plane location of the double corner and combining the orbit information (resolution ratio and incident angle) of SAR images, the parameters of the ship's status, such as the length, height and the heading direction can be calculated. Table 2 shows the corresponding incident angle of ENVISAT ASAR.

| Image Position | Breadth (km) | Distance to the Substellar Point (km) | Incident Angle (°) |
|----------------|--------------|----------------------------------------|---------------------|
| IS1 | 104.8 | 187.2–292.0 | 15.0–22.9 |
| IS2 | 104.8 | 242.0–346.9 | 19.2–26.7 |
| IS3 | 81.5 | 337.2–418.7 | 26.0–31.4 |
| IS4 | 88.1 | 412.0–500.1 | 31.0–36.3 |
| IS5 | 64.2 | 490.4–554.6 | 35.8–39.4 |
| IS6 | 70.1 | 549.7–619.8 | 39.1–42.8 |
| IS7 | 56.5 | 614.7–671.1 | 42.5–45.2 |

Table 2. Corresponding incident angle range of ENVISAT ASAR

Build a space coordinates as Fig.1 with the true north (Y axis) as the reference direction. The ship length is $l$, and the width is $w$, height is $h$, azimuthal angle of the bow is $\alpha$ (the angle between the bow and the North), and then a ship model can be descripted by a set of geometric parameters as $P = [\alpha, l, w, h]$. For example, if the satellite is descending, for ENVISAT ASAR it is a right view image. Set radar incident angle as $\theta$, the azimuthal angle as $\beta$ (the angle between the satellite and the true north) and the height as $H$, which is shown as in Fig.1. According to SAR imaging mechanism, the shadow in the figure shows the ship's scattering image, which is determined by the ship's ground position, shape and the radar's scattering orientation. Assume that the ship hull scattering length is $s$ and the width is $d$. The range is

$$w = b\cos(\beta - \alpha)$$

$$b = d\tan^2\theta \tag{10}$$

$$w = d\tan^2\theta\cos(\beta - \alpha)$$

Here $w$ is the ship width, $b$ is the width of the ship at the radar range, $d$ is the scattering width of hull target，$\theta$ is the radar incident angle, $\beta$ is the radar azimuthal angle and $\alpha$ is the bow azimuthal angle. The azimuth is

$$s = l\cos(\beta - \alpha)$$

$$l = \frac{s}{\cos(\beta - \alpha)} \tag{11}$$

$$w = d\tan^2\theta\cos(\beta - \alpha)$$

Here $l$ is the ship length and $s$ is the scattering length of the ship target images at the radar range, $\beta$ is the radar azimuthal angle and $\alpha$ is the bow azimuthal angle.

According to the length of the ship ships can be divided into small-sized (<50 meters), mid-sized (50-100 meters), big-sized (100-200 meters) and extra big-sized ship (>200 meters).

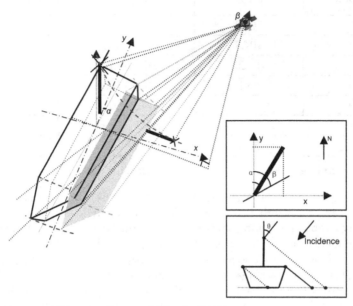

Fig. 1. The ship geometric projection model in the SAR image

## 3.3 Ship positioning correction and direction estimation

### 3.3.1 Ship positioning correction

Due to the geometric distortions of the SAR imaging, there exits measurement error between the observed position and its actual location. With this method here we can get the real position of the stern by way of detecting the ship's wake region with directional texture and calculating the convergent point of the ship's wake. Thereby, the position correcting parameter will be worked out. This method enables us to get the information of wave-making of the distressed ship when the wake profile is invisible or in violent sea status, which improves the flexibility and automaticity for understanding the marine remote sensor images.

#### 3.3.1.1 Wake region determination

A wake is the directional texture formed on the water surface immediately behind the ship. Therefore, the mean direction of the wake is consistent with the ship heading. In this algorithm, regarding the ship as the center, we divide the sea surface of the ship's neighborhood into several partially overlapped sectors. Define the angle between the edge of the sector and the positive direction of X axis as the edge direction angle of the sector, and the two edge direction angles formed by two edges of sectors are defined the direction range of the sector areas, and the median of the two edge angles is the main direction of the sector area. Calculate the textural energy of the main direction in each sector area of the ship, and set the sector area with the most textural energy as the wake areas. The direction range of the sector is regarded as the wake direction range, and the target area close to the wake range is the stern in the image.

### 3.3.1.2 Course calculation

In the wake area, calculate the textural direction pixel by pixel, and set the pixels whose textural directions belong to the wake direction interval as the wake points. Use the least-square method to calculate the mean direction of the wake, and the result is defined as the direction of the ship's wake, i.e. the course.

### 3.3.1.3 Actual lactation of the stern

Theoretically, a wake is the region of disturbed flow immediately behind the ship, and the texture of the wake pattern should converge to a point which is the real position of the stern. Collect the wake points in the wake region. Use the Hough transform to calculate the vanishing point of the wake texture. And this point is regarded as the actual location of the stern.

### 3.3.1.4 Projection offset calculation

According to the actual position of the stern and its image position, the projecting offset can be calculated. Use the stern offset, the projecting model is built. After calibration, the original detected result can be modified and the actual location of the ship target can be obtained.

## 3.3.2 Heading direction estimation method

A bow wave is the wave that forms at the bow of a ship when it moves through the water. As the bow wave spreads out, it defines the outer limits of a ship's wake. Theoretically, the convergent point of the bow wave' outline must be on the extending line of the ship's heading. The convergent point can be obtained by the Hough transform, and the heading direction of the ship can be calculated by the positions of the bow and the stern. Using the orbit information (resolution ratio and incident angle) of SAR images, the length, height and the direction of the ship can be worked out. The heading direction vector is defined in accordance with the bow wave's outline, which points from the actual location of the stern to the convergent point of the bow wave. Then, the heading direction is defined as the direction between the heading direction vector and the true north in anti-clockwise.

## 3.4 The registration algorithm of SAR image and nautical chart based on Gaussian principle curve

To implement registration to remote sensing images with navigational radar image and the chart, the detection results will be directly showed both on the remote sensing image and the chart, and then do contrast verification among the remote sensing detection results and the data of radar and AIS. A multi-scale matching algorithm of radar image and chart is proposed in this project, transforming the coastline into a set of smooth curves in the Gaussian scale space, and making coarse to fine image registration to radar image and the coastline in the chart separately in the frequency field and the spatial field.

## 3.4.1 Curve feature representation

In the extracting coastlines, many near-shore objects such as ships and navigation marks that also have strong echoes may be merged by mistakes, leading to $\Omega$-shaped spurs of the coastlines. In order to reduce the influence caused by this kind of noises, a geometric criterion is proposed to avoid selecting initial seeds on spurs. To find proper seed, each candidate seed on the rough coastline is considered by means of judging the angle between the candidate seed and its adjacent selected seed from a certain point on the land, which is the mirror of the radar image center, i.e., the own-ship position. This procedure is illustrated in Fig.2. Search for the follow-up seed to seed $v_i$ in the counter clockwise direction along the initial coastline, where $O$ is the own-ship position, and $O'$ is its mirror point perpendicular to the course. Judge the angle $\tilde{\theta}_i$ between $v_i$ and the candidate point $\tilde{v}_{i+1}$ from $O'$. Spur may occur when $\tilde{\theta}_i$ is small or even negative. Bypass those kinds of points $\tilde{v}_{i+1}$ until meet a point $v_{i+1}$ whose angle $\theta_i$ with $v_i$ is larger than a predefined threshold.

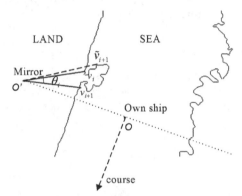

Fig. 2. Dispose the burr noise of radar coastline

Then, a family of smoothed coastlines is derived in the Gaussian scale-space, as shown in Fig.3. Scale-space is a special type of multi-scale representation that comprises a continuous scale parameter and preserves the same spatial sampling at all scales.

$$L(.;\sigma) = g(.;\sigma) * f \tag{12}$$

where the Gaussian kernel is $g_\sigma(x) = [\exp(-x^2/2\sigma^2)]/\sqrt{2\pi}\sigma$. Because the coastline in the electronic chart is rather smooth, the scale-space derivation is only done for the SAR image.

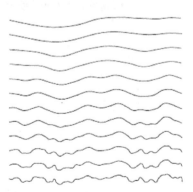

Fig. 3. Continuous smooth coastline in different scale spaces

### 3.4.2 Registration

#### 3.4.2.1 Coarse registration in the phase domain

The image registration technique based on Fourier-Mellin transform finds its applications in many different fields thanks to its high accuracy, robustness and low computational cost. It can be used to register images which are misaligned due to rotation, scaling and translation. The basic theory for translation estimation is the Fourier shift theorem. Denote

$$\mathcal{F}\{f(x,y)\} \overset{\Delta}{=} F(w_x, w_y) \tag{13}$$

which is the Fourier transform of $f(x,y)$. Then

$$\mathcal{F}\{f(x+\Delta x, y+\Delta y)\} \overset{\Delta}{=} F(w_x, w_y) e^{j(w_x \Delta x + w_y \Delta y_0)} \tag{14}$$

And the image translation can be estimated by the cross-spectrum of the two images.

$$\frac{F_1(u,v)F_2^*(u,v)}{\left|F_1(u,v)F_2^*(u,v)\right|} = e^{j(w_x \Delta x + w_y \Delta y_0)} \tag{15}$$

Assuming $s(x,y)$ is transformed image of $r(x,y)$ after translation $(\Delta x, \Delta y)$, rotation $\alpha$ and scaling $\sigma$ (in both $x$ and $y$ directions).

$$s(x,y) = r[\sigma(x\cos\alpha + y\sin\alpha) - \Delta x, \sigma(-x\sin\alpha + y\cos\alpha) - \Delta y] \tag{16}$$

And $s(x,y)$ will gain a two-dimensional pulse at the position of $(\Delta x, \Delta y)$ in the $(x,y)$ space. Then, the relation between the corresponding Fourier transform of $s(x,y)$ and $r(x,y)$ is:

$$s(u,v) = e^{-j\phi_s(u,v)} R[\sigma^{-1}(u\cos\alpha + v\sin\alpha), \sigma^{-1}(-u\sin\alpha + v\cos\alpha)] \tag{17}$$

And the corresponding amplitude spectrum is:

$$|s(u,v)| = \sigma^{-2} \left| R[\sigma^{-1}(u\cos\alpha + v\sin\alpha), \sigma^{-1}(-u\sin\alpha + v\cos\alpha)] \right| \tag{18}$$

Then, the rotation angle and the scaling factor can be calculated in the log-polar coordinates.

$$s_{p1}(\theta, \log\rho) = r_{p1}(\theta - \alpha, \log\rho - \log\sigma) \tag{19}$$

And evidently, $s_{p1}$ will gain a 2-D pulse at $(\alpha, \sigma)$ in the Hemi-Polar-Log $(\theta, \log\rho)$ coordinates. The phase-correlation method computes the transformation parameters by taking the curve as a whole, which takes the advantage of low computation cost and a good ability of noise immunity. This procedure is repeated in the Gaussian scale-space with a set of decreasing observing scales, and the two images are registered from rough to precise. And the transformation parameters are evaluated by clustering based on the evidence theory.

### 3.4.2.2 The selection of control point and registration seed

The derived curve is transformed into graph, and the weight of each node is represented by the energy defined by,

$$E_\sigma = w_i \sigma^2(v_i, v_{i-1}) = w_i \left[ \sum_{x \in v_i v_{i-1}} (C_\sigma(x) - \overline{C}_\sigma(x))^2 \right] \tag{20}$$

where $C_\sigma$ is the Gaussian curvature under scale $\sigma$ defined by the coined product of the largest and the smallest curvatures of $\overline{v_i v_{i-1}}$ .

$$C_\sigma(v_i, v_{i-1}) = L_{vv} L_w = L_{xx} L_y - 2L_x L_y L_{xy} + L_{yy} L_x \tag{21}$$

where $L_i$ is the Laplacian operator. $C_\sigma$ turns out to be a good corner detector, which is an important invariant feature to describe the structure of a derived curve in certain scale-space. And scaled energy $E_\sigma$ is a three-order vector, which describes the variance of curvature.

The nodes with big $E_\sigma$ are selected as control points. On the local straight line points the Gaussian curvature is zero, and the connections of these points form a parabolic line. Then, every two adjacent parabolic lines construct a registration curve fragment. This method assures each seed curve contain the typical topology of the local region.

### 3.4.3 Precise registration based on the principle curve graph

The Hausdoff distance is adopted as the comparability metric, and the best matching feature curve fragment is obtained by using the minimum distance classifier. This procedure is repeated in the Gaussian scale-space with a set of decreasing observing scales, and the two images are registered from coarse to fine. The Haussdoff distance between the registration curve $N_1$ and the reference curve $M_1$ is defined by Equation(22).

$$D_{Haussdoff}(N_1, M_1) = \max(d_F(N_1, M_1), d_B(N_1, M_1)) = \max_{L_j \in N_1} \min_{L_i \in M_1} \left( \left| L_i - L_j \right| \right) \tag{22}$$

If $D_{Haussdoff}(N_1, M_2) \le \varepsilon$ , then the two curves are matched. $\varepsilon$ is a given threshold. The matching metric is shown as Fig.4.

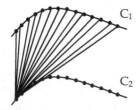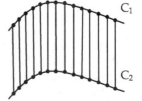

Fig. 4. The Haussdoff distance metric

### 3.4.4 Experiment analysis

The experimental images are obtained at a narrow channel in Yangzi River, China, May 1st, 2007. The own ship's position (OS POSN) is [32°13.525 N, 119°40.368 E], at speed of 13.8 knots and on course of 235.0°. The electronic chart of this area is a version in 2002. Many new docks are built, and moreover, the inland electronic chart in China uses the Gauss-Kruger coordinates, while the radar image uses the projected polar coordinates. Different coordinate systems also add extra deformation between the two images.

We choose a series of image sections from the electronic chart as the reference image, and take the radar image as image to be matched. The chart sections are selected along the coastline with a half size of the chart. And the registration is done between the chart section and the radar image. Because the two images are from different sensors, the coarse registration in single scale cannot carry out a prominent pulse in the $(\theta, \log \rho)$ space, as shown in Fig. 5. The registration procedure is repeated at scale levels $\sigma = 2^i, i = 1, \cdots, 6$. Then, the estimated transformation parameters are clustered as $(\hat{\theta}, \hat{s}, \hat{t}_x, \hat{t}_y) = [-1.9, 0.344, 353, 725]$.

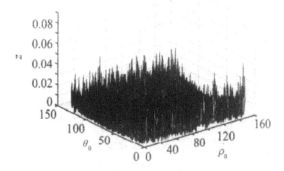

Fig. 5. The IFFT of spectrum in single scale

Twenty-one pairs of control points are selected from both the derived SAR image and the chart at the scale level of $\sigma = 16$. Using the Housdoff distance metric, the transformation parameters of second registration are obtained as $(\hat{\theta}, \hat{s}, \hat{t}_x, \hat{t}_y) = [-1.9, 0.357, 371, 660]$.

The registration results are shown in Fig.6. The registration performance is evaluated by manually registering a remote sensing image from the Google Earth with the nautical chart. The registered image is at [32°13.369 N, 119°40.279 E], 7m distance from its true position, and the rotation bias is -1.1°. The result proves that our method is feasible. Errors come from the strong echo of various objects near the shore.

### 3.5 Wave direction estimation based on local energy orientation

This method is based on Gabor filter. According to Morrone and Owens theories, local energy is the image mean square response of filter set formed by an even symmetry filter $M_e$ and an odd symmetry filter $M_o$, and it gets the biggest local energy value at singular points, such as edges and corners.

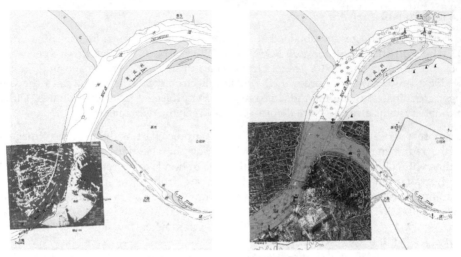

Fig. 6. Registered image pairs. (left) radar image and nautical chart, (right) remote sensing image and nautical chart

$$E(x,y) = \sqrt{(M_e * f(x,y))^2 + (M_o * f(x,y))^2} \qquad (23)$$

The steerable filter is the linear combination of a set of base filters, which are partially overlapped in the frequency domain, and can be rotated. An orthogonal filter pair is a combination of a steerable filter and its Hilbert transformation, which is designed to detect precisely the features of the edge, texture and singular point of the target. To obtain the 2-D local energy in continuous frequency space, the Wavelet Transform is used to decompose the signal into a series of sub-band signals with particular frequencies. Here we use the Mexico-hat wavelet $G_2$ to build the steerable filter $G_2^\theta$ :

$$G_2^\theta = k_1(\theta)G_2^0 + k_2(\theta)G_2^{\pi/3} + k_3(\theta)G_2^{2\pi/3} \qquad (24)$$

The Mexican-hat wavelet $G_2(x,y) = \partial^2 \exp[-(x^2 + y^2)]/\partial x^2$ is a symmetric filter with sharp narrow bandwidth, thus it can effectively restrain noise and enhance the signal in particular frequency, and it is common used in multi-scale edge detection. The fundamental filters $G_2^0$, $G_2^{\pi/3}$, $G_2^{2\pi/3}$ represent the forms of $G_2$ rotating to 0, $\pi/3$, $2\pi/3$, respectively. $k_i(\theta)$ is the interpolation function corresponding to the fundamental filters. Then, the form of $G_2$ in any orientation is represented by the linear combination of $G_2^0$, $G_2^{\pi/3}$, $G_2^{2\pi/3}$. We can get the direction energy of arbitrary pixel $(x,y)$ of the image in an arbitrary direction $\theta$ by using the orthogonal filter bank formed by steerable filter $G_2^\theta$ and its Hilbert transformation $H_2^\theta$.

$$E^\theta(x,y) = \sqrt{(G_2^\theta * f(x,y))^2 + (H_2^\theta * f(x,y))^2} \qquad (25)$$

As for the singular characteristics, e.g. the edge, when the orthogonal filter moves to the same direction with this characteristic, the direction energy reaches maximum value. The corresponding direction of the local orientation energy is called the principal direction of the pixel's local energy.

The wave image is filtered in this algorithm to eliminate speckles by way of Lee filter, and on this basis the principal energy direction of the wave can be estimated.

The experiment uses the satellite ENVISAT-1 ASAR data of 30th Sep. to 19th Oct, and the experimental area covers 30°48'N ~ 31°20'N, 122°10'E ~ 122°47'E. We use the wave direction estimation based on local energy direction to calculate the wave direction for AP polarization data. The calculation results are compared with the JMH wave analysis chart from Japan Meteorological Agency. Table 3 shows the experiment result of this wave

| Acquisition Time (UTC) | Image | Polarization Mode | Wave Direction | Direction Energy | Incident Angle | Estimated Direction | JMH Wave Analysis |
|---|---|---|---|---|---|---|---|
| 2008-09-30 13:53 | | VV | 0° | 124.3043 | 41.1016 | 75.1857 | |
| | | | 30 | 117.9377 | | | |
| | | | 60 | 132.5371 | | | |
| | | | 90 | 135.8590 | | | |
| | | | 120 | 131.9720 | | | |
| | | | 150 | 118.6072 | | | |
| | | VH | 0 | 111.2229 | | | |
| | | | 30 | 116.8233 | | | |
| | | | 60 | 133.8221 | | | |
| | | | 90 | 136.7364 | | | |
| | | | 120 | 132.8555 | | | |
| | | | 150 | 116.2534 | | | |
| 20081008 01:50 | | HH | 0 | 78.0928 | 33.9364 | 104.2056 | |
| | | | 30 | 97.9058 | | | |
| | | | 60 | 130.5053 | | | |
| | | | 90 | 145.1742 | | | |
| | | | 120 | 130.5707 | | | |
| | | | 150 | 99.4210 | | | |
| | | VV | 0 | 77.5663 | | | |
| | | | 30 | 99.3430 | | | |
| | | | 60 | 133.5955 | | | |
| | | | 90 | 146.8708 | | | |
| | | | 120 | 133.2834 | | | |
| | | | 150 | 99.9747 | | | |
| 20081010 13:39 | | HH | 0 | 14.0134 | 19.2636 | 99.9323 | |
| | | | 30 | 33.8942 | | | |
| | | | 60 | 133.1682 | | | |
| | | | 90 | 268.4664 | | | |
| | | | 120 | 132.8741 | | | |
| | | | 150 | 34.1923 | | | |

| Acquisition Time (UTC) | Image | Polarization Mode | Wave Direction | Direction Energy | Incident Angle | Estimated Direction | JMH Wave Analysis |
|---|---|---|---|---|---|---|---|
|  |  | HV | 0 | 7.7873 |  |  |  |
|  |  |  | 30 | 19.1000 |  |  |  |
|  |  |  | 60 | 74.9717 |  |  |  |
|  |  |  | 90 | 153.2269 |  |  |  |
|  |  |  | 120 | 75.1100 |  |  |  |
|  |  |  | 150 | 18.9022 |  |  |  |
| 20081019 13:56 |  | HH | 0 | 120.3329 | 44.0092 | 104.8161 | |
|  |  |  | 30 | 115.1276 |  |  |  |
|  |  |  | 60 | 120.3069 |  |  |  |
|  |  |  | 90 | 116.8428 |  |  |  |
|  |  |  | 120 | 120.2216 |  |  |  |
|  |  |  | 150 | 118.1973 |  |  |  |
|  |  | HV | 0 | 66.6512 |  |  |  |
|  |  |  | 30 | 69.6033 |  |  |  |
|  |  |  | 60 | 73.6283 |  |  |  |
|  |  |  | 90 | 71.8446 |  |  |  |
|  |  |  | 120 | 73.2198 |  |  |  |
|  |  |  | 150 | 69.0945 |  |  |  |

Table 3. The experiment result of the wave direction estimation algorithm

direction estimation algorithm. The experimental result analysis shows that VV polarization mode is the best way for wave analysis, and the following is HH, while cross polarization VH and HV mode are not ideal.

## 4. The architecture of the remote sensing aided maritime search and rescue system

The Remote Sensing Monitoring System for Maritime Search and Rescue (RS-MSR) consists of four modules including satellite transit inquiry module, vessel detection module, sea state analysis module and integrated processing module. Ship detection module has three functions and they are ship location, ship type identification/classification and ship movement direction estimation. Sea state analysis mainly estimates the wave direction. The integrated processing module receives the detection results from ship detection module and sea state analysis module. According to the distressd ownship' position, heading and the wave direction, combining the time used for data receiving, it estimates the position of the distressed ship, and combining the satellite parameter, it can revise the result obtained through ship detection. The analytic result by way of integrated processing module can be transmitted to the Maritime Safety Administration (MSA) and the rescue vessel on the working field, providing assisting decisions of areas for the rescue work. Fig. 7 describes the architecture of RS-MSR.

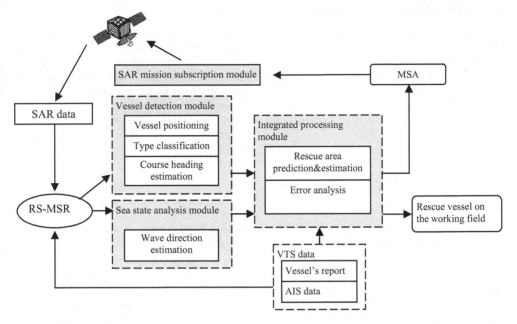

Fig. 7. The architecture of the Remote Sensing Monitoring System for Maritime Search and Rescue (RS-MSR)

## 4.1 Satellite transit inquiry module

With the development of astronavigation, the number of satellites installed with SAR sensors is increasing. Facing with so many satellites with different purposes, it has become a tough problem for clients to judge and select what they want quickly. RS-MSR sets up a real-time satellite coverage inquiry system including the commonly used satellites around the world, such as RadarSat, Envisat, ERS, CosmoSAR, TerraSAR, helping clients to retrieve quickly the crossing time and the orbit data of these satellites at specific area.

## 4.2 Ship detection module

Ship detection module is the core unit of the whole RS-MSR system. Using the micro-area images of distressed areas supplied by way of satellite, it can detect and monitor the ships and the accidental areas, supplying clue of the distressed ship for search and rescue and helping to determine the areas quickly. This module consists three parts including: (1) ship detection; (2) ship classification/identification; and (3) ship direction and course estimation.

## 4.3 Sea state analysis module

Sea state analysis module can perform initial analysis on the situation of the distressed area by estimating the wave direction and supply foundation for search and rescue decision, which is useful for estimating the floating direction and location of the distressed ships.

#### 4.4 Integrated processing module

The integrated processing module is mainly used for follow up processes of the detection result and search and rescue assisting forecast. It has two main functions. Firstly, search and rescue range estimation. According to the current location, navigation direction and the wave direction, estimate the potential searching areas of the distressed ship under the settled speed of the ship and velocity of flow. Secondly, ship location correction. Combine the satellite parameter to revise the error caused by slant-range projection imaging of the SAR images. The analysis data obtained by way of integrated processing module can be saved as *.dat or *.mat form and transmitted to marine department and the rescue spot, supplying assistant for the rescue areas determination.

### 5. System experimental performance

#### 5.1 Experiment design

In our experiment, the *Yangtze River* estuary (Changjiangkou) precautionary (30°48'N ~ 31°20'N, 122°10'E ~ 122°47'E) is selected as the experimental working zone (Fig. 8). This region includes the inward and outward fairway of the *Yangtze River* estuary and the 1# and 2# anchorages. The *Yangtze River* estuary is a tide-coordinated region, and the tide rise and fall twice per day. The tidal range is up to 4 meters. And the flow is rapid. Both anchored ships and underway ships are aggregated at this place. Casualties happen often, and it is a key region for the monitoring of maritime search and rescue. Therefore, this place is an excellent experimental region.

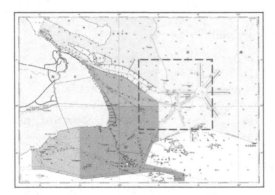

Fig. 8. The experimental working zone in the nautical chart

#### 5.2 Experiment data

From 2008.09.30 to 2008.10.19, we took four times of experiments. EnviSat-1 ASAR data is used in this experiment. Data acquisition is implemented considering the three tidal conditions: rise, fall and still. And the data includes: VTS maritime Radar, AIS, weather information, tide and flow, etc. The details are listed in Table 4. Here we present the experiment on 2008.10.19 as an example.

- Experiment region: 30°48'N~31°20'N, 122°10'E~122°47'E
- Experiment time: from 2008.09.30 to 2008.10.19

- Tide reference: Jigujiao *tide*-gauge station (31°10'24″ N, 122°22'54″ E)
- Flow reference: Xinkaihe *tide*-gauge station (31°14'36″ N, 121°29'12″ E)

| No. | Experiment Time (UTC) | Tide reference (Jigujiao) | Flow reference (Xinkaihe) |
|---|---|---|---|
| 1 | 2008-09-30, 13:53:24 | 14:00 L, 431cm | 10:00 L, 105cm/s |
| 2 | 2008-10-08, 01:50:45 | 02:00 L, 244cm | 10:00 L, 105cm/s |
| 3 | 2008-10-10, 13:39:00 | 14:00 L, 303cm | 22:00 L, -076cm/s |
| 4 | 2008-10-19, 13:56:17 | 14:00 L, 246cm | 22:00 L, 077cm/s |

Table 4. The hydrological information in the experiment

### 5.2.1 The remote sensing data

Data: ASA_APP_1PNBEI20081019_135613_000000202073_00082_34705_4038.N1
Polarization Mode: HH/HV
The detail information is listed in Table 5.

| Mode | Track | Frame | Lower Left longitude | Upper Right longitude | Upper Left latitude | Lower Right latitude | Swath | Passing direction | Start Date /Time (UTC) |
|---|---|---|---|---|---|---|---|---|---|
| APP_H H/HV | 82 | 596 | 122.167 | 122.921 | 31.6478 | 30.7285 | S7 | Descending | 2008-10-19 13:56:18.802 |

Table 5. The detail information of the remote sensing data

### 5.2.2 The VTS reference data

The radar data and AIS data are received from the VTS center of Shanghai port. Fig. 9 is the VTS shore-based radar detection picture simultaneously at the acquisition time of the remote sensing data. The VTS report shows the name and position of all the vessels equipped with AIS in the Yangtze River estuary. The Radar data and AIS data provide a reference for the performance analysis of the system experiment.

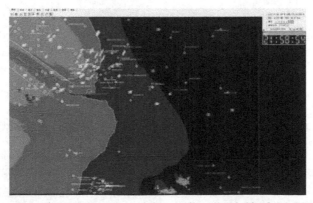

Fig. 9. The VTS report at the passing time of the satellite (2008-10-19, 13:56:17 UTC)

### 5.2.3 The weather information

The weather information includes the JMH weather chart, wave height, flow rate, and the tidal data. Fig. 10 shows the track of the No.15 typhoon (30 September, 2008).

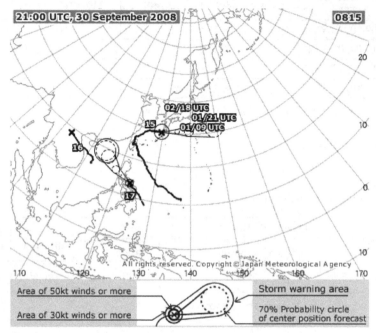

Fig. 10. The track of the No.15 typhoon (30 September,2008)

### 5.3 System experiment

Here we set detecting zone as follow:

- UpperLeft longitude: 122.338E
- UpperLeft latitude: 30.9261N
- Range in longitude: 9.8276 nmile
- Range in latitude: 8.96046 nmile

Fig. 11 presents the working interface of the Remote Sensing Monitoring System for Maritime Search & Rescue (RS-MSR). And Fig. 12 shows the original SAR image.

### 5.3.1 Vessel detection

Vessel detection includes: ship position detection, type identification, length estimation and heading estimation. The detection results are shown in Fig. 13(middle) and Table 6. The ship type classified into four categories: small, middle, large, and extra-large. The ship heading direction is the angle between the ship heading and the real north in clockwise.

In Fig. 13(right), the red circle represents small ship (point target) and the green arrow indicates the estimated heading direction of big ship (distributed corners).

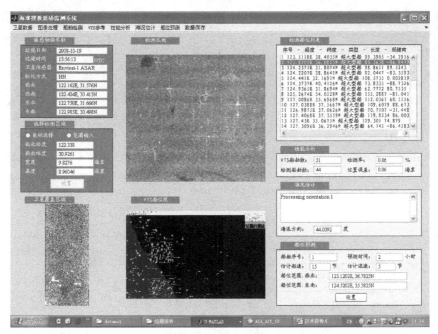

Fig. 11. The working interface of the Remote Sensing Monitoring System for Maritime Search & Rescue (RS-MSR)

Fig. 12. The original SAR image (2008-10-19, Changjiangkou precautionary, Envisat-1 ASAR HH)

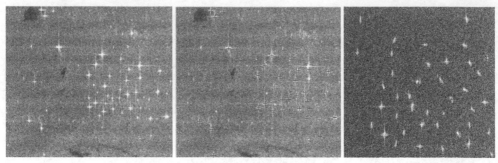

Fig. 13. (left) The selected detecting zone (original image), (middle) The vessel detection results (the red cross), (right) Ship size identification and the heading direction estimation result

| No. | Position | Type | Length(m) | Heading angle (degree) |
|-----|----------|------|-----------|------------------------|
| 1 | 123.5118E 38.4035N | M | 99.3905 | 124.2916 |
| 2 | 123.8202E 36.0825N | M | 80.3638 | 176.8479 |
| 3 | 124.2371E 31.8074N | M | 98.8651 | 0.4757 |
| 4 | 124.2207E 38.8641N | M | 92.0447 | 173.3593 |
| 5 | 124.441E 32.1691N | M | 108.5735 | 89.9972 |
| 6 | 129.429E 35.3644N | M | 95.7427 | 233.4018 |
| 7 | 129.5839E 33.1727N | L | 140.0645 | 10.0805 |
| 8 | 129.8325E 36.7182N | L | 116.6667 | 179.5634 |
| 9 | 130.0705E 35.3129N | XL | 219.6904 | 0.0432 |
| 10 | 129.9275E 38.2278N | M | 73.48 | 7.3131 |

Note: There are totally 44 vessels are detected, and only 10 are listed here.

Table 6. The vessel detection results

### 5.3.2 Performance analysis

-   VTS vessel report: 51
-   Detected vessel: 44
-   Detecting rate: 86%
-   Position error: 0.06 nmile

Among the detection, six small vessels and one middle sized vessel were missed. The performance is better in the detection of big and extra big vessels.

### 5.3.3 Sea state estimation

The wave direction estimation result is shown in Table 7.

| Acquisition Time (UTC) | Polarization Mode | Incident Angle | Estimated Direction | JMH Wave Analysis |
|---|---|---|---|---|
| 20081019 13:56 | HH/HV | 44.0092 | 104.8161 | |

Table 7. The wave direction estimation result

### 5.3.4 Ship position prediction

- Ship number: 1 #
- Predicting time: 2 hours After
- Ship speed: 15 knot
- Flow speed: 3 knot
- MSR Searching region prediction:
  - Northwest: 123.1202E, 36.7825N;
  - Southeast: 124.5202E, 35.3825N.

### 5.4 The system performance

The complete maritime search and rescue system supplies the function of ship detection, identification and location algorithm. It can be also used to inquire the satellite transit time and its orbit data. The precision of ship size estimation reached over 80% and the position estimate the position detecting error is within 0.5 nautical miles.

## 6. Conclusion

In this chapter, a remote sensing monitoring system for maritime search and rescue (RS-MSR) is presented. Some related algorithms are introduced. The satellite remote sensing imageries of large scale water area are acquired to detect and locate distress ships for guiding the search operation. Some important data such as current and sea state are retrieved to assist decision-making of the operation. System experiment design and test are presented, and the performance shows that this system can effectively improve the searching speed and accuracy, and is of significance in promoting rescue success rate and efficiency.

## 7. Acknowledgment

The research work in this paper is partially sponsored by the Shanghai Leading Academic Discipline Project (grant number: S30602), and the Natural Science Foundation of China (grant number: 40801174), and the Program of Shanghai Subject Chief Scientist (grant number: 10QA1403100).

## 8. References

Dare P., Dowman I. (2001). An Improved Model for Automatic Feature-based Registration of SAR and SPOT Images, *ISPRS Journal of Photogrammetry & Remote Sensing*, 2001, Vol.56, pp.13-28

Gerkacg J. (1999). Spatially Distributed Target Detection in Non-Gaussian Clutter. *IEEE Trans. on AES*, 1999, Vol.35, No.3, pp.926-934

Huang D.S & Han Y.Q. (1997). A Detection Method of High Resolution Radar Targets Based on Position Correction. *Journal of Electronics*, 1997, Vol.19, No.5, pp.584-590.

Kapoor R.; Banerjee A.; Tsihrintzis G.A. & et al. (1999). UWB Radar Detection of Target in Foliage using Alpha-stable Clutter Models. *IEEE Trans. on AES*, 1999, Vol.35, No.3, pp.819-833

Kuttikkad S. & Chllappa R. (1994). Non-Gaussian CFAR Techniques for Target Detection in High Resolution SAR Images. *IEEE, International Radar Conference 1994, pp.910-914*, Austin, TX, USA, 1994

Morrone, M.C. & Owens, R.A. (1987). Feature Detection from Local Energy, *Pattern Recognition Letters*, Vol. 6, pp. 303-313

Novak L.M. & Hesse S.R. (1993). Optimal Polarization for Radar Detection and Recognition of Targets in Clutter, *IEEE, National Radar Conference*, pp.79-83, Lynnfield, MA, USA, 1993

Steger, C. (1996). An Unbiased Detector of Curvilinear Structures, *Technical Report FGBV-96-03*, Forschungsgruppe Bildverstehen (FG BV), Informatik IX, Technische Universität München

Uratsuka S. & et al. (2002). High-resolution Dual-bands Interferometric and Polarimetric Airborne SAR (Pi-SAR) and Its Applications, *Proceedings of IGARSS'02*, Vol.3, pp.24-28

# Section 2

# Human Activity Assessment

# 7

# Remote Sensing Applications in Archaeological Research

Dimitrios D. Alexakis[1], Athos Agapiou[1],
Diofantos G. Hadjimitsis[1] and Apostolos Sarris[2]
*[1]Cyprus University of Technology, Department of Civil Engineering and Geomatics*
*[2]Foundation for Research and Technology, Institute for Mediterranean Studies,*
*Laboratory of Geophysical, Satellite Remote Sensing and Archaeoenvironment*
*[1]Cyprus*
*[2]Greece*

## 1. Introduction

The spectral capability of early satellite sensors opened new perspectives in the field of archaeological research. The recent availability of hyperspectral and multispectral satellite imageries has established a valid and low cost alternative to aerial imagery in the field of archaeological remote sensing. The high spatial resolution and spectral capability can make the VHR satellite images a valuable data source for archaeological investigation, ranging from synoptic views to small details. Since the beginning of the 20th century, aerial photography has been used in archaeology primarily to view features on the earth's surface, which are difficult if not impossible to visualize from the ground level (Rowland and Sarris, 2006 ; Vermeulen, F. and Verhoeven, G., 2004). Archaeology is a recent application area of satellite remote sensing and features such as ancient settlements can be detected with remote sensing procedures, provided that the spatial resolution of the sensor is adequate enough to detect the features (Menze et al., 2006). A number of different satellite sensors have been employed in a variety of archaeological applications to the mapping of subsurface remains and the management and protection of archaeological sites (Liu et al., 2003). The advantage of satellite imagery over aerial photography is the greater spectral range, due to the capabilities of the various on-board sensors.

Most satellite multi-spectral sensors have the ability to capture data within the visible and non-visible spectrum, encompassing a portion of the ultraviolet region, the visible, and the IR region, enabling a more comprehensive analysis (Paulidis, L., 2005). Multispectral imagery such as Landsat or ASTER is considered to be a standard means for the classification of ground cover and soil types (Fowler M.J.F., 2002). Concerning the detection of settlement mounds the above sensors have been proved to be helpful for the identification of un-vegetated and eroded sites. In recent years the high spatial resolution imageries of IKONOS and Quickbird have been used for the detection of settlements and shallow depth monuments (De Laet et al., 2007; 36 Massini et al., 2007; Sarris, A., 2005). Hyperspectral imagery (both airborne and satellite) has been also applied in archaeological investigations on an experimental basis and need further investigation (Cavalli et al., 2008; Merola et al., 2006).

The record of electromagnetic radiation can be achieved using special sensors. Such kind of sensors are used to record the electromagnetic radiation from satellites while handheld sensors can be used for field measurements. The ground radiometry and spectroscopy involves the study of the spectral characteristics of objects according to their physical properties (Milton, 1987). Indeed, data from portable radiometer are often refer in the literature as "ground truth data", due to the fact that measurements are collected in a relatively short distance from the object so that any noise is minimized (Jonhson, 2006). However, as Curran and Williamson (1986) emphasizes even these ground "true" data are subject to errors, which researchers should take into account.

The spectral signature diagram, from different materials or objects, is an easy way to plot radiation against wavelength, in a graphical form. Curves of spectral signature (reflectance curves) and the so-called critical spectral bands (critical spectral regions) are used in many applications of Remote Sensing (e.g. vegetation indices). The way of how measurements are collected by radiometers can be explained through physical laws. Already, by the 1970s Nicodemous et al. (1977) have proposed the basis for the model of "bidirectional reflectance distribution function" (BRDF), which describes the relationship of the incident radiation from a given address in the reflected radiation in another direction. Nevertheless, the use of the results of the Nicodemous et al. (1977) study was not appreciated and understood by the scientific community (Schaepman-Strub et al., 2006; Milton et al., 2009). Their study has been used several years later by Martonchik et al. (2000) and Schaepman-Strub et al. (2006). The original classification proposed by Nicodemous et al. (1977), depending on the geometry of the radiation which included nine categories was reduced to only four which are actually encountered (Martonchik et al., 2000; Milton et al., 2009).

Milton et al. (2009) stated that all spectroscopy measurements in a strict physical sense can be categorized within the "hemispherical-conical reflectance function" (HCRF case). It should also be noted that natural materials do not follow the rules of a diffuse Lambertian surface, since the intensity of the reflected radiation varies regarding the angle of refraction.

Ground spectroradiometer may be used to provide calibrated measurements, since these instruments are often accompanied by special Lambertian targets. Milton et al. (2009) emphasizes that a critical factor for good results is the calibration of a specific target. The only disadvantage, apart from the price of handheld spectroradiometer is that it is difficult to cover a large area (such as an archaeological site) (Atkinson et al. 1992; Milton et al., 2009).

Apart from the purchase of the ground radiometry, there is still an important limitation that should be taken into account. Most spectroradiometers which are found in the market are "single-beam": the same instrument used to measure the radiation to a specific target (reference panel) used to measure the targets of interest (target). In the interval of these measurements the atmospheric conditions are assumed to be the same.

Spectroradiometer may be used for archaeological research in order to retrieve characteristics of vegetation and to calculate vegetation indices. Such indices are quantitative measures, based on vegetation spectral properties that attempt to measure biomass or vegetative vigor. Theoretical analyses and field studies have shown that VIs are near-linearly related to photosynthetically active radiation absorbed by plant canopy, and therefore to light-dependent physiological processes, such as photosynthesis, occurring in the upper canopy (Glenn et al. 2008).

Concerning geophysical techniques they offer a non-invasive way of providing valuable information regarding the subsurface context of the archaeological sites and contribute significantly in leading archaeological excavations, reconstructing the past landscapes, suggesting directives for the cultural heritage management and preservation of sites and historical buildings and providing a prior strategy for large construction works. Employing a suite of methods measuring the physical properties of the soils with high efficiency, reliability and resolution, geophysical prospection has been designated as a valuable tool in the domain of archaeological research, especially in the study of the ancient landscapes. A very detailed review of the physical properties of each method and the fundamentals of the operation of the corresponding instrumentation is provided by Linford (2006) and Scollar et al. (1990).

Methods that involved measurements of the soil's electrical resistance and of the local intensity of the magnetic field of the earth have been the earliest that were applied in the field of archaeological prospection. Making use of Ohm's law, soil resistance meters, acting as active prospection methods, introduce a current within the ground through the use of metal electrodes and measure the electrical resistance of the soil, as this is influenced by the features located within the ground and below the electrode array (Clark, 1990). Soil resistance measurements can be carried out in two main ways, either by moving a fixed electrode spacing (corresponding to a specific penetration depth) array along profiles or through vertical electrical soundings (VES) (Sarris, 2008). In the latter, the current electrode spacing is increased with respect to a fixed centre of the electrode array, providing plots of the apparent resistivity versus electrode spacing that are ultimately compared to theoretical curves to provide information about the layering of the subsoil's strata. Soil resistance values are modified depending of the resistivity contrast of the targets (eg. a high resistance wall structure or a high conductive ditch) with the surrounding soil matrix. The measured apparent resistivity is also affected by the type of the configuration of electrodes (e.g. Twin probe, Square, Wenner, pole-pole, dipole-pole) and the distance between the electrodes, which is relatively proportional to the penetration depth.

This chapter seeks to address applications of remote sensing and GIS in archaeological research in a three-fold way. Initially, potential of satellite remote sensing is highlighted through a multi – sensor case study in Thessaly, Greece, where different satellite image processing techniques contributed to the detection of Neolithic tells (the so called 'magoules') that are found in the Thessaly plain. Four satellite remote sensing images with different spatial resolutions (ASTER, Landsat, HYPERION, IKONOS) were examined in order to search their potential for automatic extraction of Neolithic settlements, by means of pixel – based (RGB composites, spatial and radiometric enhancement, vegetation indices, data fusion, classification methods, data fusion etc). The satellite data were statistically analyzed, together with other environmental parameters, to examine any kind of correlation between environmental, archaeological and satellite data. Moreover, different methods were compared and integrated methodologies for the detection of Neolithic settlements were extracted.

Concerning ground spectroradiometer contribution to archaeological research, new innovative tools and methodologies are also presented in this chapter. Specifically, ground "truth" data, presented as spectral signatures libraries, were provided from different spectro-radiometric campaigns at archaeological environments (e.g. Tombs of the Kings and Sikyon archaeological site, C. Greece). Moreover field spectroscopy was used to detect buried archaeological sites in similar ways as applied in remote sensing applications in Neolithic tells in Thessaly. In addition the comparison of the phenological cycle 5 profile of similar crops - under same meteorological and soil conditions - is also searched over archaeological and non archaeological sites concerning for different case studies in Cyprus.

At the end, in order to highlight the potential of geophysical remote sensing in archaeological research certain case studies from surveys held in Greece and Europe are presented such as magnetic surveys in the Neolithic settlement of Veszto –Bikeri in Hungary and the Byzantine walls of ancient Nicopolis in Greece, the ground penetrating radar methods in the ancient Agora of Feres (Velestino) in Thessaly, in Agora of Sikyon at NE Peloponesse - Greece and in the area of the hypothesized amphitheatre of Ierapetra (SE Crete and finally the electrical resistivity tomographies from the Nemea, Peloponesse.

## 2. Application of remote sensing, GIS and geomorphology to the reconstruction of habitation in Neolithic Thessaly

### 2.1 Introduction

The aim of this study is to highlight the contribution of different approaches such as Remote Sensing, GIS and geomorphology analysis for the detection of Neolithic settlements and the modeling of habitation in the area of Thessaly - Greece.

The Neolithic settlement mud mounds in the area of Thessaly, Greece are called Magoules. They are low hills of 1-5 meters height and mean diameter 300 meters. The vast majority of Magoules are laid on Larisa basin and a smaller number is distributed in west Thessaly (Karditsa basin) (Fig.1). Both of these plains consist of Quaternary alluvial deposits (Alexakis et al. 2008). In order to achieve the goals of the research it was necessary to proceed with the topographic mapping of the settlements through the use of GPS, digitize 1:50,000 scale topographic and geological maps and construct a detailed archaeological and environmental database in SQL environment.

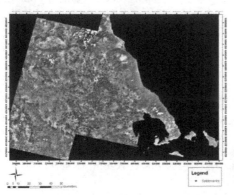 

Fig. 1. Color composite RGB→3,2,1 of the mosaic of the 4 ASTER images used to cover the whole area of interest. The dots represent the location of the Neolithic magoules (left). Magoula Aerino (right).

## 2.2 Landscape reconstruction

Prior to the landscape reconstruction of Thessaly during different Neolithic periods, the reliability of the existing Digital Elevation Model (DEM) was evaluated compared to other digital elevation models, such as the 90m pixel size DEM (from the Shuttle Radar Topography Mission (SRTM) and the 30m pixel size DEMs provided by ASTER images or constructed by the L1-A stereoscopic products (3N and 3B) of ASTER. The results indicated that the RMSE for the DEM created through the digitization of the contour lines of the topographic maps was the lowest. The two major plains of Thessaly contain 181 out of the 342 known registered "magoules", stressing the important role of reconstruction of the relief of each basin during each Neolithic sub-period. Both geological (stratigraphic data from boreholes and past geomorphologic studies) and archaeological data were placed under consideration for achieving this task and a reconstructed DEM for each basin and each Neolithic period was created.

## 2.3 Satellite image processing

### 2.3.1 Data and preprocessing of satellite images

Concerning this study different multispectral images were used. Specifically, 4 IKONOS images of 1m pixel size, 1 Landsat ETM+ 30m pixel size image, one 30 m pixel size Hyperion image and 4 ASTER images (15, 30 and 90m pixel size). Masking of the sea, clouds and snow areas in all images preceded the processing of the images in order to focus to the mainland and the areas that provided useful information. Image mosaics were created accordingly depending on the types of sensors and both image mosaics and isolated images were rectified to a common projection system (EGSA87/HGSR87). Digital numbers of images were also converted to reflectance values according to specific conversion equations. The last step was necessary in order to have a uniformity in the values of images originating from different sensors.

### 2.3.2 Spectral enhancement techniques

Several RGB composites were constructed in an effort to examine their efficiency in the detection of the Neolithic settlements. For the ASTER image with acquisition date 19-03-2003, where most of the magoules were registered, the RGB→1,2,3, RGB→3,2,5 and RGB→2,3,7 composites were the most successful for the visual detection of the Neolithic settlements (39 out of 239 settlements were highly visible, 49 average visible and 151 poorly visible). All these composites appeared to have the highest Optimum Index Factor (Alexakis et al, 2009). Similarly, RGB composites of IKONOS images were able to detect 27 out of 48 settlements within their area of coverage. It is worth mentioning that 19 of the detectable magoules, namely the highest of all corresponding to an average altitude of 4.6m, were highly visible in all RGB composites. On the other hand, RGB composites of Landsat and HYPERION images were not very promising (for HYPERION composites only 5 out of 21 settlements were detected). Due to their high spatial resolution, all the 5 settlements that felt within the spatial limits of the aerial photo mosaic were easily detectable. As a general conclusion however, the acquisition date of the images proved to be the most crucial factor for the detection of magoules mainly due to the intensive cultivation (mainly soft and shallow cultivation) of the landscape both on the top and the surroundings of magoules. Principal Component Analysis (PCA) was applied to ASTER, Landsat and Hyperion images, being especially effective for ASTER images where 39 and 47 out of 247 settlements were highly or medium discriminated correspondingly. Image fusion techniques through the combination of high spatial resolution images such as IKONOS (1m) and high spectral resolution images such as Hyperion (30m) concluded to very promising results (Fig 2). Finally, a spectral mixer utility (Erdas Imagine 9.1 software) contributed to exploit the dynamic range of all the multispectral information of the Hyperion image by combining more than three bands to an RGB composite. Using the specific utility and assigning a weighting coefficient for each band, a RGB composite of 23 bands (38, 42, 48, 49, 50, 51, 52), (85, 86, 87, 88, 89, 90, 91, 92,) & (93, 94, 108, 109, 110, 111, 113, 114) was constructed that enhanced the visual appearance of the magoules.

### 2.3.3 Spectral enhancement techniques

Radiometric enhancement was vital for the appearance of the images. After applying radiometric enhancement to ASTER images (acquisition date of 19-03-2003) we managed to detect 57 settlements (Fig. 2). A non-linear radiometric enhancement of the HYPERION PCA image, followed by an inversion of brightness was able to highlight 8 settlements from a total of 9. (Melia 1, Melia 2, Anagennisi 2, Moshohori 3, Kipseli 2, Prodromos 1 of Larisa, Nikaia 17 and Kuparissia 2). Similar type of non-linear radiometric enhancement of the high resolution IKONOS images through the modification of the histogram outlined the round shape of known magoules, as well as outlined 10 more targets of similar geometry that need to be verified by the ground truthing activities that will follow (Fig. 2).

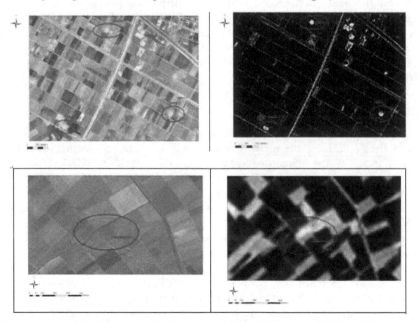

Fig. 2. Appearance of 3 settlements in the original IKONOS image (left) and the radiometrically enhanced image where three Neolithic settlements are highlighted (right). To the north of Galini-3 settlement, shown at the lower right of the image), another smaller potential magoula is suggested.

### 2.3.4 Land classification, vegetation indices, spatial enhancement

A spectral signature database was constructed to provide the basic spectral information about tells, especially at the plain areas of Thessaly. Several classification methods were applied to Landsat and ASTER images in order to investigate the land use regime around the magoules. Examination of the overall accuracy of the various algorithms tested (based on the error matrix), proved that the Mahalanobis algorithm was the most efficient for the exact classification of the images (Fig. 3a). Additionally, object based segmentation techniques were applied to ASTER images and 15 settlements in total of 234 were detected easily (Alexakis, 2009). The computation of the Normalized Difference Vegetation Index (NDVI) was used to highlight the vegetation differences during different periods of time, in an effort to pinpoint any vague indications for the detection of magoules. As expected, the NDVI of the "spring" ASTER image was higher than the "summer" Landsat image, but still the vegetation differences of the spring time favored the detection of magoules mainly due to the differentiations in the soil's humidity (Fig. 3b).

Application of certain spatial high pass filters contributed further to the spatial enhancement of smaller features such as the magoules. The most reliable of them proved to be Sobel Right Diagonal 3x3 and Laplace 3x3, both of which outlined clearly the limits of the most prominent of them (Fig. 3c).

Fig. 3. (a) Land classification of ASTER image through the use of Mahalanobis distance (fuzzy) algorithm. (b) Detail from the application of NDVI to ASTER image. (c) ASTER image around Halki area after the application of Sobel Right Diagonal filter. Neolithic magoules are indicated within the ellipses.

### 2.4 Analysis in GIS environment

An extensive spatial analysis of the magoules distribution was carried out in GIS environment using the reconstructed DEMs. Besides the extraction of statistics regarding the relation of settlements to the aspect, slope and relief height, the distance of settlements from natural resources was calculated by applying buffer zones around the quarries and the water springs (mainly springs existing on the mountainous areas). Watersheds were constructed and the distance of each settlement from its neighbor watershed was calculated. Density maps of the settlements were created for each Neolithic period. The calculation of the density of the settlements was accomplished through the use of a non-parametric Kernel technique. The spatial territorial limits of the settlements were explored using the Thiessen polygons analysis. The site catchment of Neolithic settlements was studied through least cost surface analysis. Cost surfaces contributed also to the exploration of communication routes between the different settlements (Alexakis, 2011).

Finally, GIS tools were employed to construct predictive habitation models for each phase of the Neolithic period in an effort to locate areas that could possibly host similar type of settlements. The specific predictive models were based on the use of a multi-parametric spatial analysis method of geographic elements and other information (statistical, archaeological, a.o.). All the environmental factors (height, aspect, slope, distance from watersheds, distance from water springs, distance from quarries, geology, viewshed, distance from chert sources, least cost paths, a.o.) that could affected the choice of habitation in Neolithic Thessaly were statistically examined and certain weight factors were applied to each one of them. At a final stage, a fuzzy logic algorithm and a normalization equation were also applied a more efficient tuning of the results and for rating the final probability from 0 to 1.

## 2.5 Application of sophisticated fields to the Digital Elevation Models

The final approach of the particular project involved the detection of Neolithic settlements through the analysis of DEMs with the use of three different semi–automated methodologies. Three different DEMs (90m pixel size SRTM DEM, 30m ASTER DEM and a 20m DEM from the digitization of contours of topographic maps) were tested in all procedures to attest for their potential in the detection of the magoules.

The first methodology involved the estimation of the index of convexity (CI) to the three different DEMs according to Fry *et al.* (2004):

$$CI = (x - x_{med}) / (x_{max} - x_{med}), \tag{1}$$

where x is the initial DEM, $x_{med}$ is the DEM after the application of median 7x7 filter and $x_{max}$ is the DEM after the application of maximum 7x7 filter.

Although the index of convexity seemed to be ideal for the detection of low hills such as magoules, only 35 (28%) of them at Larisa plain and 28 (47%) at Karditsa plain were detected by this method.

The second methodology is related with the design and application of customized filters similar to those used by Menze and Sherratt (2006). The optimal filter for the detection of a signal with a well known shape is the matched filter (Fig. 4). For the construction of the matched filter, an area of 5x5 pixel was cut around the DEM of each settlement for a total of 50 settlements. Then, the value of the central pixel was subtracted by each pixel, followed by stacking of all the 50 local DEMs (through the layer stack utility of Erdas Imagine software) to form a final multilayer image. The particular image was imposed to Principal Component Analysis and the first 5 principal components were summed. The negative sum replaced the value of the central pixel of each of the above 5 filters, which they were then applied individually to Larisa and Karditsa basins as a detection filter (Fig. 14). The statistics for these filters proved that the specific methodology is really promising especially for the SRTM DEM in the area of Larisa. More specifically, about 60% of the settlements were detected through the application of the first and second filters.

The third methodology followed the approach of Iwahashi and Kamiya (1995) for the estimation of the geometric signatures of DEM through a combined study of slope gradient, surface convexity and texture (Fig. 4). Specifically, binary files were formed through estimation of the mean value of slope gradient: all pixels with value above the mean value took a value 1 and all the rest took a value of 0. The same binary archives were created after the subtraction of the initial DEM from the one that has been processed through the application of a median filter. The last binary image was created after the application of a Laplace 3x3 filter to the initial DEM. In the end, the three binary archives were summed and the final map highlighted the areas of high local convexity. Fuzzy logic algorithms were applied to the final results of the filtered DEM in order to produce a better classification scheme (Pixel values equal to 0 formed the first group, values from 0 to 3 formed the second group and pixels with value equal to 3 formed the third group) (Fig. 15). The application of this methodology to SRTM resulted to the detection of 35% of the magoules in Larisa plain and only 15% of the magoules in Karditsa plain.

The results obtained through the Menze & Sherratt (2006) approach to the SRTM DEM were also implemented to the predictive modelling, together with other subproducts of the satellite image analysis, such as the NDVI map, land use classification and spectral signatures library of magoules. A similar methodology of significant weights and factors was considered and results were subjected to fuzzy logic and normalization techniques. Still, the results of predictive modelling did not alter significantly from the previous approach, signifying a state of saturation for the parameters considered.

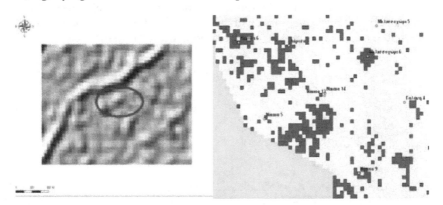

Fig. 4. Magoula Turnavos 6 after the application of matched filter (left). Application of geometric signatures methodology to SRTM DEM in the area of Larisa. With the red color are indicated the areas of higher height where magoules could be established (right).

## 3. Application of field spectroscopy and satellite remote sensing to archaeology

### 3.1 Introduction

This chapter aims to introduce the capabilities and the potentials of field spectroscopy to archaeological research (Fig. 5). Field spectroscopy involves not only the acquisition of accurate measurements (e.g. spectral signatures profiles) but also the study of the interrelationships between the spectral characteristics of objects and their biophysical attributes in their field environment. Therefore, field spectroscopy can provide valuable information for an area if we consider the fact that human eye senses only a small part of the electromagnetic spectrum, from approximately 0.4 to 0.7 nm, whereas field spectroscopy in support of remote sensing operates in a wider spectrum range including near infrared as well. In this section, ground "truth" data, presented as spectral signatures libraries, are provided from different spectroradiometric campaigns at archaeological environments (e.g. Tombs of the Kings and Nea Paphos at SW Cyprus, Sikyon archaeological site, C. Greece). Furthermore, spectral libraries include vegetation profiles, mainly over barley crops (from the Palaepaphos – Cyprus archaeological site and from Neolithic tells at Thessaly - Greece). Such libraries are used in order to examine either the seasonal changes of vegetation, or the anomalies of vegetation profiles due to buried archaeological remains. Moreover, spectral libraries are used for the atmospheric correction of satellite imagery. Finally, the theoretical background of scaling up ground narrow bands taken from handheld spectroradiometers to bandwidths satellite imagery, using the Relative Response Filters, is presented.

Ground spectroscopy may be used as a fast detection method in order to evaluate positive or negative crop marks. In this case sections over archaeological areas are taken and evaluated in terms of vegetation indices. Different Neolithic sites at Thessaly (central Greece) are examined with the use of this approach. Finally, an alternative method for the detection of archaeological remains is presented in this chapter. This method is based on the comparison of the phenological cycle profile of similar crops -under same meteorological and soil conditions -over archaeological and non archaeological sites. The case of Palaepaphos site in Cyprus is presented with the use of medium resolution images (Landsat TM/ETM+) and the support of ground spectroradiometric measurements.

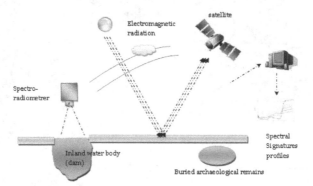

Fig. 5. Potentials of ground field spectroscopy for archaeological purposes

## 3.2 Collection of field measurements

Field measurements were carried out in different archaeological areas in Cyprus and Greece. The areas investigated and presented in this chapter were fully vegetated with crops. Moreover ground spectroradiometric measurements were taken: a) over visible monuments in order to develop an archaeological spectral signature database and b) at dams (inland clear water) for applying atmospheric correction to satellite images.

The spectroradiometric instrument that was used to register the spectral signature was GER 1500 (Fig. 6). This instrument may record electromagnetic radiation from a range of 350 nm up to 1050 nm. It includes more than 500 different channels and each channel cover a range of about 1.5 nm. The field of view (FOV) of the instrument was set to $4^\circ$ ($\approx 0{,}02$ m$^2$).

Fig. 6. GER 1500 used in this study with its calibration target (Agapiou et al. 2010)

A reference spectralon panel was used to measure the incoming solar radiation. The Labertian spectralon panel (≈100% reflectance) measurement was used as references while the measurement over vegetated areas or archaeological sites as a target. Therefore reflectance for each measurement can be calculated using the following equation (2):

$$\text{Reflectance} = (\text{Target Radiance} / \text{Panel Radiance}) \times \text{Calibration of the panel} \qquad (2)$$

In order to examine the use of broadband vegetation indices such as NDVI, narrow band reflectance (from the spectroradiometer) needed to be recalculated according to the spectral characteristics of a specific satellite sensor. The authors selected to simulate these data to Landsat TM /ETM+ satellite imagery based on Relative Spectral Response (RSR) filters. RSR filters describe the instrument relative sensitivity to radiance at various part of the electromagnetic spectrum (Wu et al. 2010). These spectral responses have a value of 0 to 1 and have no units since they are relative to the peak response (Fig. 7, left). Bandpass filters are used in the same way in spectroradiometers in order to transmit a certain wavelength band, and block others. The reflectance from the spectroradiometer was calculated based on the wavelength of each sensor and the RSR filter as follows:

$$R_{band} = \Sigma\ (R_i * RSR_i)/\ \Sigma RSR_i \qquad (3)$$

Where: $R_{band}$ = reflectance at a range of wavelength (e.g. Band 1)
$R_i$ = reflectance at a specific wavelength (e.g $R$ 450 nm)
$RSR_i$ = Relative Response value at the specific wavelength

To avoid any errors due to significant changes in the prevailing atmospheric conditions, the measurements over the panel and the target are taken in a short time. In this case it is assumed that irradiance had not significant change which is true for non hazy days (Milton et al. 2009). Finally the measurements were carried out between 10:00 and 14:00 (local time) in order to minimize the impact of illumination changes on the spectral responses (Milton, 1987) at a height of 1.2 m (Fig. 7, right)

### 3.3 Spectral libraries

Spectral signature diagram is an easy way to plot target reflectance against wavelength, in a graphical form. Therefore ground field measurements from archaeological sites may be used in order to create an "archaeological" digital spectral signature library. Even though several remote sensing applications investigate the correlation of the spectral signature of an object,

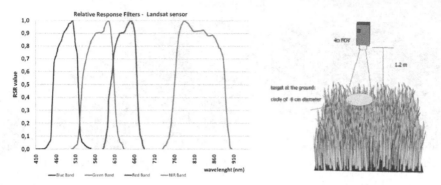

Fig. 7. Relative Response filters for Bands 1-4 of Landat TM sensor (left) and typical diagram of the in-situ spectroradiometric measurements.

in the majority these applications the aim is exactly the opposite: the study and identification of "unknown" targets through the spectral signature. Therefore "archaeological" spectral libraries may be used for identification or correlation of different archaeological sites remotely.

Different spectral signatures from the archaeological site "Tombs of the Kings" (Agapiou et al. 2011a) and "Sikyon" archaeological sites were taken. Spectral profiles indicate that there is great potential for detecting archaeological remains in the spectral range between 550 – to 850 nm (from the green visible part of spectrum to near infrared) because of the extremely different spectral response of the archaeological material compared to sand and local marl/carbonate sandstone in the archaeological site "Tombs of the Kings"(Fig. 8). Similar results were found for "Sikyona" site also (Fig. 9).

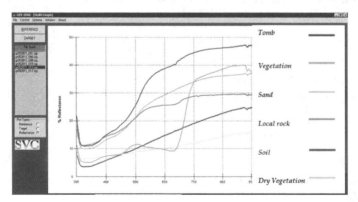

Fig. 8. Spectral signatures profiles from different targets at the archaeological site "Tombs of the Kings".

Spectral signatures libraries proved to be really efficient for any potential researcher that may use satellite imagery in order to detect archaeological relics in the area because it highlights the high correlation of spectral response of archaeological material, sand and local geological formations in the area of red visible band.

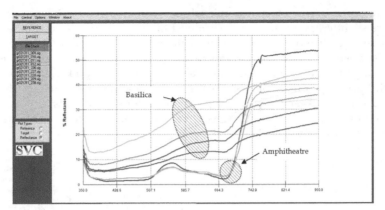

Fig. 9. Spectral signatures from the archaeological site of Sikyon.

## 3.4 Atmospheric correction of satellite images based on ground data

Atmospheric correction of satellite images is a necessary pre-processing step before any image analysis. Earth's surface radiation, undergoes significant interaction with the atmosphere before it reaches the satellite sensor. This interaction is stronger when the target surfaces consist of non-bright objects, such as vegetated areas examined in many archaeological studies. This problem is especially significant when using multi-temporal satellite data for monitoring purposes (Hadjimitsis et al. 2010). As Lillesand et al. (2004) argue satellite images need to be atmospherically corrected before being subjected to any post-processing techniques.

Atmospheric effects are a result of molecular scattering and absorption of the incoming radiation and influence the quality of the information extracted from remote sensing images. Such errors occurred by atmospheric effects can increase the uncertainty up to 10%, depending on the spectral channel (Che and Price, 1992). Hadjimitsis et al. (2010) have also highlighted the importance of considering atmospheric effects when several vegetation indices, such as NDVI were applied to Landsat TM/ETM+ images for agricultural applications. In their study a mean difference of 18% for the NDVI was recorded before and after the application of darkest pixel method. Therefore atmospheric correction is an important pre-processing step required in many remote sensing applications since is needed to convert the at-satellite spectral radiances of satellite imagery to their at-surface counterparts.

The modified Darkest Pixel (DP) atmospheric correction method (Hadjimitsis et al., 2004) was applied to multi-series Landsat images (Agapiou et al., 2011b). The surface radiance of the dark targets is assumed to have approximated zero surface radiance or reflectance. Instead of assuming $L_{darkest\ target}$ to be zero value, the modified DP considers the 'true' ground radiance or reflectance value over dark targets as the $L_{darkest\ target}$.

For this reason ground spectroradiometric measurements were taken in inland clear water target (Asprokremmos Dam in Paphos). The GER-1500 field spectroradiometer was equipped with a fibre optic probe was used in order to retrieve the spectral signatures from the dam. After using the RSR filters for Landsat TM/ETM+ images the spectral reflectance after the atmospheric correction was calcualated. The results have shown that satellite images were slightly improved after the removal of atmospheric effects. Indeed crop and soil marks from archaeological areas were enhanced. Photo interpretation quality was enchanced at images with low water vapour optical thickness and in general for images with water vapour optical thickness less than 0.05, the quality of the images after the atmospheric correction was improved. In the case of higher values the quality was not improved sufficiently (Agapiou, 2011a). Fig. 10 shows some typical histograms before and after the atmospheric correction. As it is shown the initial histogram of the image is stretched and therefore interpretation is improved.

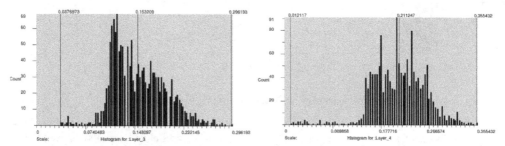

Fig. 10. Histogram for Band 3 before (left) and after (right) atmospheric correction (image: Landsat TM, 25-09-2009).

Generally the interpretation showed that in cloud-free image with low water vapour optical thickness ($\approx$ < 0.05) atmospheric correction can increase the quality of the satellite image and therefore improve the interpretation.

## 3.5 Verification of buried archaeological sites

Field spectroscopy can be also used for detection of buried archaeological remains. The advantage of using ground spectroscopy against satellite remote sensing, is based on the fact that the researcher may repeatable use such methodology in contrast to the temporal resolution of satellite images (e.g. 16 days for Landsat images). Although spatial resolution is increased (few cm) the extent and scale of spectroscopy is limited compared to the area coverage of a single satellite image.

For the verification of known archaeological sites using ground spectroscopy, GER 1500 spectroradiometer was used in several vegetated archaeological sites. In this chapter results from field campaigns over Neolithic tells in Thessaly (central Greece) and buried remains in the Palaepaphos area (SW Cyprus) are presented (Agapiou et al., 2010; Agapiou and Hadjimitsis,2011 ; Agapiou et al., 2012a). In each archaeological site several sections were carried out. For the first site, along each section, more than 50 ground spectroradiometric measurements were taken while in each consecutive 5th measurement the calibration spectralon panel was used in order to minimize sun changes illuminations. At the second case study measurements were taken over known geophysical anomalies (potential subsurface monuments). To avoid differences due to variations in cultivation techniques, all measurements were carried out within the same parcel. As it shown in Fig. 11, vegetation indices such NDVI and Simple Ratio tend to give higher values at the highest peak of the tell, similar to other flat – healthy crops of the area, while the slope of the tell gives lowest NDVI and SR values. This is due to the fact that top of the tell seems to have similar hydrological behaviour as the flat healthy region (e.g. same level of water surface run off and similar inclination ≈ 0%) in contrast to the slope of the tell. The sloping part of the tell seems to behave differently due to rainfall erosion processes. All these results denote the correlation between the morphology and the spectral response of canopy on the magoules (Agapiou et al., 2012a). Moreover ground spectral signatures at Palaepaphos area (Fig. 11) indicated a stress condition for crops over the geophysical anomaly in contrast to the rest of the measurements. This stress condition was detected from ground spectroradiometric measurements as shown in Fig. 11.

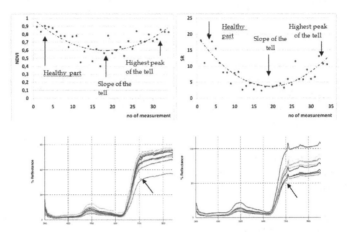

Fig. 11. NDVI (Top left) and Simple Ratio (Top right) profile over archaeological site at Thessaly. Spectral signature profiles over geophysical anomaly (indicated with arrow) at Palaepaphos (Bottom).

The results of Thessaly were able to be confirmed using Landsat TM/ETM+ images. Indeed as it was found the similar characteristics were observed and in satellite images (Fig. 12). Therefore using this experience of the spectroradiometer, where ground hyperspectral data were collected, a researcher focusing in satellite imagery can seek and search for similarly spectral characteristics as those in the spectroradiometric campaign.

## 3.6 Monitoring phenological cycle of crops

Monitoring the phenological cycle of crops for archaeological sites has been very limited discussed in the literature. Nevertheless as Agapiou and Hadjimitsis (2011) argue, that this approach may be used –under some assumptions- for the detection of buried archaeological remains. This methodology may be used in cases were spatial resolution of satellite imagery is very low or the cost of high multispectral satellite imagery is forbidden for an archaeological research. The basic theory of the applied method is based on the different spectral signature characteristics of 'stressed' (negative crop mark due to buried walls) and 'non-stressed' (i.e. healthy) vegetation based on the following two criteria: (a) similar soil characteristics & (b) similar climatic characteristics. The determination of spectral signatures of barley can be also verified using field spectro-radiometric measurements.

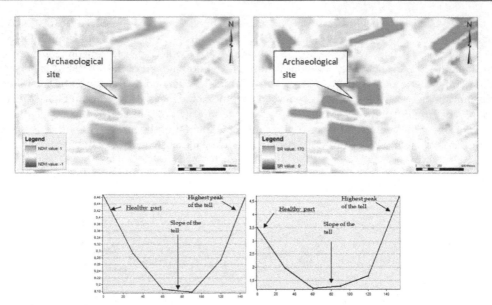

Fig. 12. NDVI (Top left) and Simple Ratio (Top Right) images over archaeological site at Thessaly using Landsat TM image. Characteristics sections of NDVI (Bottom left) and Simple Ratio (Botom right) over the tell Nikaia 6.

In their study Agapiou and Hadjimitsis (2011) and Agapiou et al. (2012b) have used fifteen Landsat TM and ETM+ images all freely available from USGS Glovis database. After applying the necessary pre-processing steps, such as geometric and radiometric corrections, the NDVI algorithm was applied in three selected case studies where barley crop was cultivated. The whole phenological cycle of barley crops was examined for a period of one year, from June 2009 until June 2010, using Landsat TM/ ETM+ images, in order to detect areas of "possible" archaeological remains indicated as spectral signatures anomalies. Site 1 was an archaeological area excavated in July 2010 by the Department of Antiquities, while sites 2 and 3 were healthy sites. Moreover site 3 was in close proximity to site 1 in order to minimize errors due to different climatic or soil characteristics.

At the same time meteorological data have not shown any significant variations over these sites (temperature, precipitation and humidity). Fig.13 shows the red and near infrared values during the phenological cycle. As it is expected in a healthy situation (similar to the Tasseled Cap algorithm, see Kauth and Thomas 1976) after the first rains the vegetation starts to grow until its reach to its highest peak (see Site 2, Fig.13). However this is not the case for stress crops as in the case of the archaeological site (site 1). A stress condition is indicated as it is shown in Fig.13 (Point D) which may be related to the presence of archaeological remains in the area. Fig.13 presents the phenological cycle of the three sites as examined by Agapiou and Hadjimitsis (2011). As indicated in 07/01/2010 an immediate drop of NDVI value was found for Site 1 (archaeological site). The low NDVI value could be explained as a result of the presence of areas of potential archaeological site, which affected the growth of the crop. The agricultural barley crop in Site 1 can be characterized as a "stressed" vegetation (negative crop mark predominantly found above walls). The excavations carried out in the area have

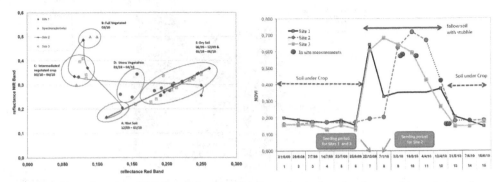

Fig. 13. Red band against NIR band during the phenological cycle of crops (Agapiou and Hadjimitsis, 2011) (left). NDVI for archaeological and non archaeological areas. In situ Spectroradiometric measurements are plotted as dots (right).

verified that this localized crop stress was due to the presence of archaeological remains (walls). Positive crop marks, due to ditches (crop vigour), were not found either in the methodology applied or in the excavated area. In the non-stress area of Site 3, crops were still growing this period indicating a peak of the NDVI value.Crops in Site 2 have not yet grown at this time. From 07/01/2010 until 31/05/2010, crops in Site 2 start to grow gradually until the harvest period. The chronological shift (X – axis) occurred in NDVI peaks for Sites 1, 3 (at 22/12/2009) and Site 2 (at 19/03/2010), as recorded from the satellite images, were due to different land management of the fields in these two areas.

## 4. Ground based geophysical techniques

Electrical resistivity tomography (ERT) uses various combinations of distances among equally spaced electrodes (of a specific array) in order to extract information about the lateral and vertical variations of the apparent resistivity.

Magnetic methods on the other hand are passive, measuring the magnetic properties of the soils or the magnitude of the local magnetic field of the earth as it is modified by the magnetic properties of the underlying features. The local magnetic field of the earth is modified by either the enhancement of iron oxides due to past activities or due to the burning of features, which upon cooling they keep a permanent thermoremanent magnetization, quite distinct from the current magnetic field (in intensity and direction) (Aitken, 1974 ; Sarris & Jones, 2000). Based on this property, magnetometers measure either the total magnetic field (through proton precession or cesium magnetometers) or one of its components, for detecting variations of the magnetic field (magnetic anomalies) that can be caused by anthropogenic buried agents (Nishimura, 2001). In order to account for the diurnal variations of the magnetic field and counterbalance effects resulting from geological trends, these instruments were also used in a gradient mode. Similarly, the use of fluxgate gradiometers (measuring the vertical magnetic gradient) was essential for increasing the sensitivity (although not comparable to that of the cesium or optically pumped alkali vapor magnetometers) and sampling pace of the magnetic surveys (Fig. 14). Penetration depth of the magnetic gradient surveys depends on the distance between the two vertical magnetic sensors.

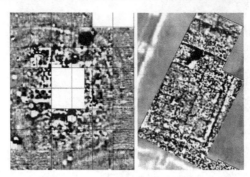

Fig. 14. Results of the usage of Geoscan FM256 and Bartington G601 fluxgate gradiometers for the mapping of the Early Copper Age (ca. 4,500-3,900 BC) settlement of Veszto-Bikeri, in the Great Hungarian Plain (left image) and the urban center of ancient Nikopolis (Epiros, NW Greece) (right image). The magnetic survey on the Hungarian Bronze Age tell revealed three circular ditches encircling the settlement, which consists of a dense cluster of structural remains (rectangular houses, pits, kilns, etc) (Sarris et al. 2004). In the case of ancient Nikopolis, a large complex building (45x70m) with a number of internal divisions appeared right at the edge of the SW corner of the Byzantine Paleochristian walls (Sarris et al., 2010).

Electromagnetic methods (EM), including the ground penetrating radar (GPR) and the soil conductivity techniques (SC), have been also employed for the prospection of archaeological sites and the reconstruction of the ancient terrains. The Slingram type of soil conductivity devices (such as Geonics EM31 or EM38) operate at low frequencies (usually at the range of 50-300kHz), make use of the electromagnetic induction and are capable of providing measurements of both soil's apparent electrical conductivity (quadrature component) and apparent magnetic susceptibility (in-phase component), with various penetration depths depending on the orientation (vertical coplanar (VCP) and horizontal coplanar (HCP) orientations), the frequency of operation and separation of the transmitter and receiver coils (Dalan, 2006; Gaffney & Gater, 2003; Cheethman, 2010). The strength of EM signals that are registered by the receiver system depends on the conductivity of the soils, the magnetic permeability and the dielectric permittivity (especially for the GPR). Operating within the range of radio frequencies, GPR systems consist of a transmitter antenna that sends a signal (~30-1000MHz) which propagates through the different strata or features (reflectors) of the subsurface and a receiver antenna that registers all the secondary reflections (with a modified amplitude) that arrive to it after a time delay which is converted to the depth. Signal attenuation and penetration depth of the GPR decreases with the increase of the frequency of the antennas and the conductivity of the soils. GPR signals are collected with high sampling rate along transects and the resulting reflection sections (radargrams) represent the variation of the amplitude of the reflected signal with depth and thus they depict an image of the stratigraphy of the subsurface. GPR parallel transects are usually combined to created 3D volumetric maps of the subsurface and through the isolation of specific time or depth slices it is possible to allocate the horizontal extent of archaeological features at different depths, which is of importance especially in cases that one wants to have information regarding the vertical extent of the features or to construct 3D models of them. In this sense, the GPR survey can be valuable in mapping different occupation strata and resolving features that are located at various depths (Fig. 15) (Conyers, 2004; Conyers & Goodman, 1997).

(a)            (b)              (c)                (d)

(a) Section of a Stoa from the ancient Agora of Feres (Velestino) in Thessaly. The (image dimensions: ExN=16x40m).
(b) Monumental structural remains SE of the Zeus temple in the archaeological site of Nemea, Peloponesse. (image dimensions: ExN=7x15m) (Papadopoulos *et al.* 2011).
(c) Architectural complex at a depth of 90-100cm below the surface located to the east of the Agora of Sikyon at NE Peloponesse (image dimensions: ExN=30x50m).
(d) The 2.5-3m GPR depth slice from the area of the hypothesized amphitheatre of Ierapetra (SE Crete). The survey was carried out in the area that was suggested through the rectification of the map of the British Vice-Admiral Thomas Spratt (1811-1888) (which was depicting the approximate location of the amphitheatre) on the satellite (Quickbird) image of the region. The deeper GPR slices in combination to ERT measurements provided evidence for the underlying relics of the amphitheatre (Sarris et al., 2011).

Fig. 15. Examples of GPR time slices obtained by Nogin Plus (Sensors & Software) GPR system using a 250MHz antenna.

Even microgravity measurements have been carried out for the detection of features that have a substantial mass density contrast with respect to their surrounding geological domain, creating a difference at the local gravitational acceleration. Measurements of earth's gravitational acceleration are carried out though the use of gravimeters that measure the acceleration of gravity within a hundredth of a mGal ($1Gal=1cm/sec^2$) or less. As such, the resolution of the method is dependent on the size and volume of the targets and requires tedious corrections and processing (such as drift correction, latitude correction, free air correction, Bouguer correction, etc) as measurements are influenced by the regional or even local trends. A recent review of archaeological, environmental and geological microgravity applications has been provided recently by Eppelbaum (2011).

Mainly used for landscape reconstruction, large monumental structure detection and deep prospection surveys, seismic techniques exploit acoustical waves generated either by a sledge hammer or an explosion discharge. In seismic refraction, the acoustic wave sensors (geophones) are laid along specific distances and record the refracted signals with respect to their arrival times and in this way their velocity of propagation (increasing with depth) is measured. Seismic reflection techniques require a smaller distance between the source and the geophones and through the examination of the arrival times, amplitude and shape of the reflected waves, we can conclude on the types of the subsurface interfaces (Metwaly et al, 2005 ; Scott & Markiewich, 1990).

If the above techniques are capable of providing a mean of detection and localization of architectural features within an archaeological site, magnetic susceptibility (MS) measurements and chemical analyses can contribute in providing a further tool for investigating the land use patterns at a specific area. Magnetic susceptibility provides not only a measure of the effectiveness of the potential application of magnetic surveys (through the estimation of the normalized Le Borgne Contrast, namely the variation of the magnetic susceptibility with depth), but also an index of the past workshop activities in an area. Measurements of the magnetic susceptibility and the frequency dependent susceptibility (namely the variation of MS with the frequency of an induced magnetic field) are capable in distinguishing soils enriched in single domain magnetic particles (from the geological origin multidomain particles) which are indicative of the intensity of the occupation of a site (Clark, 1990; Mullins & Tite, 1973; Thompson & Oldfield,1986). Coupled with results of chemical properties of soils (especially those dealing with phosphate analysis or heavy metals tracing) it is possible to characterize the type of workshop activities (eg. increase of manganese content can be associated to glass workshop activities) or differentiate areas used for animal husbandry, midden deposits, foundation trenches, cultivation, cooking, etc. Even chemical stability of certain organic chemical compounds (e.g. coprastanol) may act as a biomarker of the human presence at a particular locatio (Sarris, 2008).

The choice of the technique depends mainly on a number of factors: the type of the targets, their lateral and vertical dimensions, their deposition depth and type/properties of the surrounding soils (to be able to create a significant signal, contrast or "anomaly"). Architectural features such as stone/brick structures, roads, walls, built/chamber or rock-cut tombs, can be relatively easily resolved through soil resistance or GPR surveys. Brick structures or architectural features that are either burnt or contain residues of heating/burning, kilns, workshop facilities, slag deposits, metal concentrations, and sometimes roads, walls and fortifications can be detected through magnetic and electromagnetic techniques. The use of ERT, GPR and microgravity is especially useful for the identification of vaults, caves, chamber tombs and fissures. Shallow depth surveys usually employ magnetic, soil resistance techniques and the use of GPR. In cases where deeper penetration is required GPR, ERT and seismic approaches are more appropriate (Fig. 3) (Sarris 2008, Linford 2006).

Ground based prospection techniques are not only limited to the survey of archaeological sites in an open/rural context (Fig. 16). They can also be applied within an urbanized environment, but in such a case only specific techniques can be used (such as ERT and GPR) that are influenced as less as possible by the modern interventions and structures that exist in the urban matrix. A lot of these applications do not only involve the mapping of the subsurface (eg. below asphalt roads, pavements, concrete blocks, etc), but sometimes they are oriented towards the stability or structural damage assessment of monuments or historical structures aiming towards their architectural restoration (Bertroli et al., 2011; Pettinelli et al., 2011; Utsi, 2010; Masini et al., 2010).

Currently there are two different tendencies in archaeological prospection: the integration of different geophysical techniques for maximizing the information content and the employment of multi-sensor methods for the rapid coverage of sites. In most cases, the integrated use of various techniques is employed to extract more information about an archaeological site, allowing the interpretation of various measurements that are dealing with different properties of the soil. The fusion of this information permits a more holistic approach as the data can complement each other and provide a more integral plan of the subsurface features. One the other hand, the recent development of new multi-sensor (for magnetics), multi-antennas (for GPR) or multi-electrode (for soil resistivity) motorized systems carrying DGPS allow the fast and detailed assessment of large regions, although

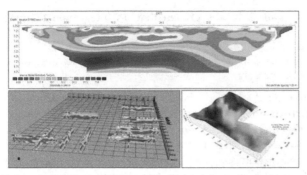

Fig. 16. Example of the application of ERT in the area of the assumed hippodrome in the archaeological site of Nemea, Peloponesse. A number of ERT transects was materialized reaching the depth of about 9m below the surface. Upon the synthesis of all the 2D resistivity inverted sections (a), it was possible to create a scatter plot of all the sections (b) and through interpolation techniques the 3D volumetric resistivity map. The isolation of the various strata of the subsurface was based on the range of their resistivity values (c). In the particular case, the iso-resistivity surfaces that resulted from the ERT transects did not identify any specific leveling of the subsurface at the west side of the archaeological site where the hippodrome was expected, suggesting that the original hypothesis of the archaeologists has to be rejected (Papadopoulos et al., 2011).

they are restricted by the surface coverage and the terrain morphology (Linford et al., 2011; Doneus et al., 2011). Although the particular systems offer increased sampling density and rates of coverage, they often suffer from positioning errors due to high measuring velocities and the introduction of noise due to the non uniform balancing of the sensors or multichannel GPR systems (Zollner et al., 2011; Verdonck & Vermeulen , 2011).

Finally, image processing techniques play a significant role in the visualization of the results of the geophysical surveys as the ultimate goal is to provide images that they depict the underlying features at their exact location and horizontal/vertical extent in a way that can approach the results of an after-the-excavation plan. This objective can be achieved through the use of a number of filtering/convolution processes, the employment of synthetic models or inversion algorithms, or other image processing functions (Papadopoulos & Sarris, 2011; Sarris, 2008; Loke & Barker, 1996; Scott & Markiewich, 1990). In cases that multiple datasets are available for the same region, composites can be made using visualization techniques similar to those used in satellite remote sensing (Böniger & Tronicke, 2010). Even more impressive visualization can be created through the fusion of geophysical data with satellite remote sensing or aero-photogrammetric data and lidar or terrestrial 3D laser scanning (Bem et al., 2011). Indeed, the continuous improvement of high resolution satellite remote sensing sensors has made possible their simultaneous utilization with conventional geophysical data affecting their resolution and potential in the detection and mapping of underground features (Crespi et al., 2011).

## 5. Conclusions

The various approaches applied on different satellite images for the detection of Neolithic settlements in Thessaly illustrated the benefits that satellite remote sensing can provide in archaeological investigation. It was proven that an integration of images from different satellite sensors can contribute to a faster and more accurate and qualitative detection of archaeological sites. In addition, the GIS spatial analysis and DEM processing contributed substantially to the detection and monitoring of settlements and modeling of Neolithic habitation partners in Thessaly.

Moreover, it was proved that spectroradiometric measurements can be used as an alternative approach in order to identify buried archaeological remains, since they can provide accurate spectral signatures for a wide spectral region. Anomalies of the crop spectral signatures due to buried archaeological remains can be recorded in detail and contribute to the construction of a predictive archaeological model in the future. However, the real benefit of this instrument is when it is used in conjunction with satellite images. Moreover the spectroradiometric measurements highlight the high correlation of spectral response of archaeological material, sand and local geological formations in the area of red visible band. Finally it was proved that the monitoring of the phenological cycle of crops can be used for the detection of buried archaeological sites.

At the end geophysical survey, ground and space remote sensing methods have been gradually adopted in archaeological research in an effort to capture the residues of the past anthropogenic activity underlying below the current surface of the ground and to provide a more synthetic and holistic image of the archaeological landscapes. Magnetic and resistivity techniques, electromagnetic, gravity and seismic methods, measurements of the chemical and magnetic properties of the ground, have been all mobilized to produce an accurate picture of the underlying monuments, contributing to a variety of applications in both urban and rural environmental settings. In this way all these different methodologies Have been applied to various areas of potential archaeological interest throughout Europe. The tuning of the methods and the corresponding instrumentation, together with the development of specific processing algorithms, are necessary in order to enhance the shallow depth signals that are registered within the increased noise levels of upper horizons of the soil. In this way, shallow depth prospection techniques have been used to map architectural relics, to guide excavations, and to identify craft, workshop, agricultural or animal husbandry activities.

## 6. Acknowledgments

Part of this chapter is part of Dr. Dimitrios Alexakis Phd thesis. This research project was co-financed by INSTAP (Institute for Aegean Prehistory) and PENED (E.U.-European Social Fund (75%) and the Greek Ministry of Development-GSRT (25%))."Field Spectroradiometer and Archaeology" chapter results are part of the PhD thesis of Mr. Athos Agapiou. The authors would like to express their appreciation to the Alexander Onassis Foundation for funding the PhD study. Thanks are given to the archaeologists Dr. K. Vouzaxakis and Dr. J. Lolos, for their valuable assistance during field campaigns at the Neolithic tells (Thessaly) and the archaeological site of Sikyona, Corinthos and to the Department of Antiquities of Cyprus for their permission to carry out filed measurements at different archaeological sites of Cyprus. The whole project has also been co-financed by the internal programs "Integration" and "Monitoring"of Cyprus University of Technology. Finally thanks are given to the Remote Sensing Laboratory of the Department of Civil Engineering & Geomatics at the Cyprus University of Technology for the support (http://www.cut.ac.cy).

## 7. References

Agapiou, A., Hadjimitsis, D.G., Alexakis, D., Sarris, A. (2012a). Observatory validation of Neolithic tells ("Magoules") in the Thessalian plain, central Greece, using hyperspectral spectroradiometric data. *Journal of Archaeological Science*, 39 (5): 1499-1512, 10.1016/j.jas.2012.01.001.

Agapiou, A., Hadjimitsis, D.G., Alexakis, D., Papadavid G. (2012b). Examining the Phenological Cycle of Barley (Hordeum vulgare) for the detection of buried archaeological remains: the case studies of the Thessalian plain (Greece) and the Alampra (Cyprus) archaeological test fields. GIScience and Remote Sensing (in press).

Agapiou A.and Hadjimitsis D. G.,(2011). Vegetation indices and field spectro-radiometric measurements for validation of buried architectural remains: verification under area surveyed with geophysical campaigns, Journal of Applied Remote Sensing, 5: 053554-1.

Agapiou A.; Alexakis D.; Hadjimitsis D.G. & Themistocleous K. (2011a). Earth Observations and Ground Measurements for Cultural Heritage Protection: the Case Study of Cyprus, *1st International Conference on Safety and Crisis Management in the Construction, SME and Tourism Sectors*, June 24th – 28th 2011, Nicosia, Cyprus (in press).

Agapiou A., Hadjimitsis D. G., Papoutsa C., Alexakis D. D., Papadavid, G. (2011b). The Importance of Accounting for Atmospheric Effects in the Application of NDVI and Interpretation of Satellite Imagery Supporting Archaeological Research: The Case Studies of Palaepaphos and Nea Paphos Sites in Cyprus. Remote Sensing, 3: 2605-2629, doi: 10.3390/rs3122605.

Agapiou A.; Hadjimitsis G. D.; Sarris A.; Themistocleous K. & Papadavid G. (2010). Hyperspectral ground truth data for the detection of buried architectural remains, *Lecture Notes in Computer Science* 6436, pp. 318–331, 2010.

Aitken, M. (1974). *Physics and Archaeology*, 2nd ed., Oxford: Clarendon Press.

Alexakis, D., Astaras, Th., Sarris, A., Vouzaxakis, K., Karimali, L., (2008). Reconstructing the Neolithic Landscape of Thessaly through a GIS and Geological Approach. In: Polsluchn, K. Lambers and I. Herzog (eds.), *Layers of Perception. Proceedings of the 35th International Conference on Computer Applications and QuantitativeMethods in Archaeology/(CAA)*, Berlin, Germany.

Alexakis, D., Sarris, A., Astaras, T., Albanakis, K., (2011). Integrated GIS, remote sensing and geomorphologic approaches for the reconstruction of the landscape habitation of Thessaly during the neolithic period. *Journal of Archaeological Science* 38, 89-100.

Alexakis, D.; Sarris, A.; Astaras, T.;Albanakis, K. (2009). Detection of Neolithic settlements in Thessaly (Greece) through multispectral and Hyperspectral satellite Imagery. Sensors 9 (2), 1167- 1187. doi:10.3390/s90201167.

Atkinson P. M.; Webster R. & Curran P. J., (1992). Cokriging with Ground-Based Radiometry, *Remote Sensing of Environment*, 41, pp. 45–60.

Bartroli, S. R., Garsia, G. E. & Tamba, R. (2011). GPR Imaging of Structural Elements. Case Study of the Restoration Project of the Modernist Historic Site of Sant Pau, in Archaeological Prospection - Extended Abstracts, ed. by M. G. Drahor & M. A. Berge, 9th *International Conference of Archaeological Prospection*, Izmir, Turkey, pp. 149-151.

Bem, C.; Bem, C; Asandulesei, A.; Venedict, B. & Cotiuga, V., (2011). Identity in Diversity, in Archaeological Prospection - Extended Abstracts, ed. by M. G. Drahor & M. A. Berge, 9th *International Conference of Archaeological Prospection*, Izmir, Turkey, pp. 25-28.

Böniger, U. & Tronicke, J. (2010). Integrated data analysis at an archaeological site: A case study using 3D GPR, magnetic, and high-resolution topographic data, *Geophysics*, Vol. 75, No. 4, pp. B169-B176.

Cavalli, R.M.; Colosi, F.; Palombo, A; Pignatti, S. &Poscolieri, M. (2007).Remote Hyperspectral Imageryas a support to Archaeological Prospection, *J. Cult. Herit., 8*, 272-283.

Che, N. & Price, J. C. (1992) Survey of Radiometric calibration results and methods for visible and near infrared channels of NOAA-7, -9, and -11 AVHRRs. *Remote Sensing of Environment*, 41, 19 – 27.

Cheetham, P. (2010). An Empirical Reassessment Of The Utility Of The Geonics Em38B, Together With Suggested Methodologies For Its Application In Archaeological Investigations, *Meeting of the New Surface Geophysics Group - Recent Work in Archaeological Geophysics*, London.

Clark, A. J. (1990). *Seeing Beneath the Soil. Prospecting Methods in Archaeology*. London: B.T. Batsford Ltd.

Conghe, S. & Woodcock, E. C. (2003) Monitoring Forest Succession With Multitemporal Landsat Images: Factors of Uncertainty, *IEEE Transactions on Geoscience and Remote Sensing*, 41 (11), 2557 -2567.

Conyers, B. L. & Goodman, D. (1997). *Ground Penetrating Radar: An Introduction for Archaeologists*, Walnut Creek, CA., Altamira Press.

Conyers, L B. (2004). *Ground Penetrating Radar for Archaeology*, Walnut Creek, CA: AltaMira Press.

Crespi, M.; Dore, N.; Partuno, J.; Piro, S. & Zamuner, D. (2011). Comparison of SAR data, Optical Satellite Images and GPR Investigations for Archaeological Site Detection, in Archaeological Prospection - Extended Abstracts, ed. by M. G. Drahor & M. A. Berge, 9th *International Conference of Archaeological Prospection*, Izmir, Turkey, pp. 169-173.

Curran P. J. & Williamson H. D., (1986). Sample Size for Ground and Remotely Sensed Data, *Remote Sensing of Environment*, 20, pp. 31–41.

Dalan, R.A. (2006). Magnetic Susceptibility in Johnson, J.K. (ed.) *Remote Sensing in Archaeology: An Explicitly North American Perspective*, Alabama Press: Tuscaloosa, 162-203.

De Laet, V.; Paulissen, E. & Waelkens, M. (2007). Methods for the extraction of archaeological featuresfrom very high-resolution IKONOS-2 remote sensing imagery, Hisar (southwest Turkey). *J.Archeol. Sci., 34*, 830–841.

Demitrack, A. (1986). The Late Quaternary Geologic History of the Larisa Plain,Thessaly, Greece: Tectonic, Climatic and Human Impact on the Landscape, Ph.D.thesis, Stanford University.

Doneus, N.; Flory, S.; Hinterleitner, A.; Kastowsky. K.; Kucera, M.; Nau, E.; Neubauer, W.; Scherzer, D.; Schreg, R.; Trinks, I.; Wallner, M. & Sitz, T. (2011). Integrative Archaeological Prospection - Case Study Stubersheimer Alb. Bridging the Gap Between Geophysical Prospection and Archaeological Interpretation, in Archaeological Prospection - Extended Abstracts, ed. by M. G. Drahor & M. A. Berge, 9th *International Conference of Archaeological Prospection*, Izmir, Turkey, pp. 166-168.

Eppelbaum, V. L. 2011. Review of Environmental and Geological Microgravity Applications and Feasibility of Its Employment at Archaeological Sites in Israel, *International Journal of Geophysics*, vol. 2011, issue 1, pp. 1-9.

Fowler, M.J.F. (2002). Satellite Remote Sensing and Archaeology: a Comparative Study of Satellite Imagery of the Environs of Figsbury Ring, Wiltshire. *Archaeol.Prospec.*, *9*, 55-69.

Fry, G.L.A.; Skar, B.; Jerpansen, G.; Bakkestuen, V. & Erikstad, L., (2004). Locating archaeological sites in the landscape: a hierarchical approach based on landscape indicators. Landscape and Urban Planning 67, 97-107.

Gaffney, C. & Gater, J. (2003). *Revealing the Buried Past*, Tempus: Stroud.

Glenn E. P.; Huete A. R. ; Nagler P. L. & Nelson S. G., (2008). Relationship Between Remotely-sensed Vegetation Indices, Canopy Attributes and Plant Physiological Processes: What Vegetation Indices Can and Cannot Tell Us About the Landscape. *Sensors*, 8, 2136-2160.

Hadjimitsis, D.G.; Papadavid G.; Agapiou A..; Themistocleous K.; Hadjimitsis M. G.; Retalis A.; Michaelides S.; Chrysoulakis N.; Toulios L. & Clayton C. R. I. (2010). Atmospheric correction for satellite remotely sensed data intended for agricultural applications: impact on vegetation indices, *Nat. Hazards Earth Syst. Sci.*, 10, 89-95 .

Hadjimitsis, D.G.; Clayton, C.R.I. & Hope, V.S. (2004). An assessment of the effectiveness of atmospheric correction algorithms through the remote sensing of some reservoirs. *International Journal of Remote Sensing.*, 25, 3651-3674.

Iwahashi, J., & Kamiya, I., (1995). Landform classification using digital elevation model by the skills of image procesingdmainly using the Digital National Land Information. *Geoinformatics* 6 (2), 97 - 108 (in Japanese with English abstract).

Jonhson J. K., (2006). *Remote Sensing in Archaeology*, The University of Alabama Press, Tuscaloosa.

Kauth R. J. & Thomas G. S. (1976). The tasseled Cap - A Graphic Description of the Spectral-Temporal Development of Agricultural Crops as Seen by LANDSAT". In Proceedings of the Symposium on Machine Processing of Remotely Sensed Data, Purdue University of West Lafayette, Indiana, 4B,44-51.

Lillesand, T. M.; Kiefer, R. W. & Chipman, J. W. (2004). *Remote Sensing and image interpretation*, Wiley International Edition.

Linford, N. (2006). The Application Of Geophysical Methods To Archaeological Prospection. *Reports on Progress in Physics*, v.69, pp. 2205–2257.

Linford, N.; Linford, P.; Payne. A.; David, A.; Martin, L. & Sala, J. (2011). Stonehenge: Recent Results from a Ground Penetrating Radar Survey of the Monument, in Archaeological Prospection - Extended Abstracts, ed. by M. G. Drahor & M. A. Berge, 9th International Conference of Archaeological Prospection, Izmir, Turkey, pp. 86-89.

Loke, M. H. & Barker, R. D. (1996). Rapid Least-Squares Inversion of Apparent Resistivity Pseudo-Sections using Quasi-Newton method. Geophysical Prospecting, v.48, pp. 181-152.

Martonchik, J. V.; Bruegge, C. J. & Strahler, A. H. (2000). A review of reflectance nomenclature used in remote sensing. Remote Sensing Reviews, 19, pp. 9–20.

Masini, N. & Lasaponara, R. (2007). Investigating the spectral capability of Quickbird data to detectarchaeological remains buried under vegetated and not vegetated areas, J. Cult. Herit., 8, 53-60.

Masini, N.; Persico, R. & Rizzo, E. (2010). Some Examples Of GPR Prospecting For Monitoring Of The Monumental Heritage, Journal Of Geophysics And Engineering, vol. 7, pp. 190-199.

Menze, B.H. & Sherratt, A.G. (2006). Detection of Ancient Settlement Mounds: Archaeological SurveyBased on the SRTM Terrain Model. Photogramm. Eng. Remote Sens., 72, 321-327.

Merola, P.; Allegrini, A.; Guglierra, D. & Sampieri, S. (2006). Buried Archaeological Structures Detection Using MIVIS Hyperspectral Airborne Data, In Proceedings of SPIE, the International Society for Optical Engineering, 2006; pp. 62970Z.1-62970Z.12.

Metwaly, M.; Green, A. G.; Horstmeyer, H.; Maurer, H.; Abbas, A. M. & Hassaneen, A. G. (2005). Combined Seismic Tomographic And Ultrashallow Seismic Reflection Study Of An Early Dynastic Mastaba, Saqqara, Egypt. Archaeological Prospection, 12, pp. 245–56.

Milton E. J. & Rollin E. M. (2006). Estimating the irradiance spectrum from measurments in a limited number of spectral bands, Remote Sensing of Environment 100, pp. 348-355.

Milton, E. J. (1987). Principles of Field Spectroscopy. Remote Sensing of Environment 8 (12), 1807--1827.

Milton, E. J.; Schaepman, M. E.; Anderson, K.; Kneubühler, M & Fox, N. (2009). Progress in Field Spectroscopy. Remote Sensing of Environment 113, 92--109.

Mullins, C. E. & Tite, M. S. (1973). Magnetic Viscocity, Quadrature Susceptibility and Frequency dependence of Susceptibility in Single-Domain Assemblies of Magnetite and Maghaemite, Journal of Geophysical Research, no. 78, pp. 804-809.

Nicodemus, F. F.; Richmond, J. C.; Hsia, J. J.; Ginsberg, I.W. & Limperis, T. L. (1977). Geometrical considerations and nomenclature for reflectance. National Bureau of Standards Monograph, 160, Washington, D.C U.S. Govt. Printing Office.

Nishimura, Y. (2001). Geophysical prospection in Archaeology in D. R. Brothwell and A. M. Pollard (eds.), Handbook of Archaeological Sciences, pp.543-553.

Papadopoulos N.G.; Sarris, A.; Michalopoulou, S. & Salvi, M.C. (2011). Integrated Geophysical Investigations in Nemea and Tsoungiza, 9th International Conference of Archaeological Prospection, September 19 – 24, Izmir, Turkey.

Papadopoulos, N. & Sarris, A. (2011). An Algorithm for the fast 3-D Invesrion of Direct Current Resistivity Data Using LSMR, in Archaeological Prospection - Extended Abstracts, ed. by M. G. Drahor & M. A. Berge, 9th *International Conference of Archaeological Prospection,* Izmir, Turkey, pp. 197-200.

Pavlidis, L. (2005). High resolution satellite imagery for archaeological application. www.fungis.org/images/newsletter/205-1.pdf

Pettinelli, E.; Barone, P.M.; Mattei, E. & Lauro, S.E. (2011). Radio Waves Technique for Nondestructive Archaeological Investigations. *Contemporary Physics,* Taylor & Francis Group, London, Volume 52, Issue 2, pp. 121-130.

Rowlands, A. & Sarris, A. (2006). Detection of exposed and subsurface archaeological remains using multi– sensor remote sensing. *J. Archaeol. Sci.,* 34, 795-803.

Sarris, A.; Papadopoulos N.G.; Salvi, M.C. & Dederix, S. (2011). Preservation Assessment of Ancient Theatres through Integrated Geophysical Technologies, *XVI[th] Congress of the International Union of Prehistoric and Protohistoric Sciences (UISPP),* Florianopolis, Brazil.

Sarris, A.; Teichmann, M.; Seferou, P. & Kokkinou, E. (2010). Investigation of the Urban-Suburban Center of Ancient Nikopolis (Greece) through the Employment of Geophysical Prospection Techniques, Computer Applications and Quantitative Methods in Archeology *"Fusion of Cultures"* CAA'2010, Fco. Javier Melero & Pedro Cano (Editors), Granada, Spain.

Sarris, A. (2008). Remote Sensing Approaches / Geophysical, in *Encyclopedia of Archaeology,* ed. By Deborah M. Rearsall, Academic Press, New York, vol. 3, pp. 1912-1921.

Sarris, A. (2005). Use of remote sensing for archaeology: state of the art. Presented at the *International Conference on the Use of Space Technologies for the Conservation of Natural and Cultural Heritage, Campeche, Mexico.*

Sarris, A., Galaty, M. L.; Yerkes,R. W.; Parkinson, W. A.; Gyucha, A.; Billingsley, D. M. & Tate, R. (2004). Geophysical prospection and soil chemistry at the Early Copper Age settlement of Vésztó-Bikeri, Southeastern Hungary , *Journal of Archaeological Science,* volume 31, Issue 7, pp. 927-939.

Sarris, A. & Jones, R. E. (2000). Geophysical and Related Techniques Applied to Archaeological Survey in the Mediterranean: A Review, *Journal of Mediterranean Archaeology* (JMA), v.13, no.1, pp.3-75.

Schaepman-Strub, G.; Schaepman, M. E.; Painter, T. H.; Dangel, S., & Martonchik, J. V. ( 2006). Reflectance quantities in optical remote sensing-Definitions and case studies. *Remote Sensing of Environment,* 103, pp. 27–42.

Scott, J. H. & Markiewich, R. D. (1990). Dips and chips-PC programs for analyzing seismic refraction data: *Proceedings, SAGEEP 1990,* Golden, Colorado, pp. 175-200.

Thompson, R. & Oldfield, F. (1986). Environmental Magnetism. London: Allen and Unwin.

Utsi, E. (2010). Reflections From Westminster Abbey, *Meeting of the New Surface Geophysics Group - Recent Work in Archaeological Geophysics,* London.

Verdonck, L & Vermeulen, F. (2011). 3-D GPR Survey with a Modular System: Reducing Positioning Inaccuracies and Linear Noise, in Archaeological Prospection - Extended Abstracts, ed. by M. G. Drahor & M. A. Berge, 9th *International Conference of Archaeological Prospection,* Izmir, Turkey, pp. 204-212.

Vermeulen, F. & Verhoeven, G. (2004). The contribution of aerial photography and field survey to thestudy of urbanization in the Potenza valley (Picenum). *J. Roman Archaeol.*, *17*, 57-82.

Wu X., Sullivan T. J. & Heidinger K. A. (2010). Operational calibration of the Advanced Very High Resolution Radiometer (AVHRR) visible and near-infrared channels. Canadian *Journal of Remote Sensing*, 36 (5), 602–616.

Zollner, H.; Kniess, R. & Meyer, C. (2011). Efficient Large-scale Magnetic Prospection Using Multichannel Fluxgate Arrays and the New Digitizer LEA D2, in Archaeological Prospection - Extended Abstracts, ed. by M. G. Drahor & M. A. Berge, 9th *International Conference of Archaeological Prospection*, Izmir, Turkey, pp. 201-203.

# 8

# The Mapping of the Urban Growth of Kinshasa (DRC) Through High Resolution Remote Sensing Between 1995 and 2005

Kayembe wa Kayembe Matthieu[1],
Mathieu De Maeyer[2] and Eléonore Wolff[2]
*[1]University of Lubumbashi*
*[2]Université Libre de Bruxelles*
*[1]D.R. Congo*
*[2]Belgium*

## 1. Introduction

The study of urban growth in Kinshasa is not a new topic, as shown by the work on the dynamics of housing in the 1970s (Flouriot et al., 1975; Pain, 1978). These authors followed the spatial extension of Kinshasa by collecting old cartographic documents and comparing them. Flouriot (1975) combined a cartographic approach with household surveys to follow the long-term housing growth.

The advent and development of remote sensing and Geographic Information Systems (GIS) have changed the methods, making it now possible to map and quantify urban growth quickly and easily. More recently in Kinshasa, Tshibangu et al. (1997) have integrated into a GIS a vegetation map drawn by Compere in 1960. This was possible thanks to the interpretation of aerial photographs and Landsat and SPOT images conducted respectively in 1982 by Wilmet and 1987 by Nsekera to quantify the urban sprawl. Delbart and Wolff (2002) evaluated the extension of the city of Kinshasa from an old map (1969) and the delineation of the city in 1995 observed on a SPOT image (from 1995). The current extension of the city (between 1995 and 2005) is not precisely known, but the figures are around 600km$^2$ (Lelo Nzuzi, 2008). The purpose of this chapter is to map and quantify urban growth between 1995 and 2005 using a time series of high resolution satellite images.

## 2. Study area

The city of Kinshasa province, located between 4 ° and 5 ° south and between 15 ° and 17 ° east, is the largest city in the Democratic Republic of Congo. It covers an area of 9965 km$^2$ (De Saint Moulin, 2005), about 600 km$^2$ being only urbanized. The city had 400,000 inhabitants in 1960 and reached more than six million in 2008, the average annual growth rate between 1960 and 2003 would therefore be about 6.80% (Lelo Nzuzi, 2008).

Kinshasa has grown in the plains bordering the Congo River. The plain 300 metres above sea level covers about 200 km$^2$. This is the most industrialized area and formerly the most densely inhabited, commonly called the "ville basse" (low city). After independence in 1960, the city has spread into the complex hills surrounding the city and low peaks around 600 m above sea level. This area is mainly occupied by slums, called the "high".

## 3. Data

Two SPOT images dating respectively from March 31, 1995 and July 1, 2005 were used. They are recorded in panchromatic and multispectral modes. Their radiometric quality is variable. The 1995 images have a cloud cover of 7% in multispectral and 5% in panchromatic, while the 2005 images have 6% of cloud cover in multispectral mode and 10% in panchromatic mode. The presence of these clouds is evidence of the difficulty of obtaining cloud-free images for areas located in the sub-equatorial climate. To make the different images comparable, a radiometric correction was performed. Unfortunately, due to the low correlation between the red and the green bands, it did not yield good results and was abandoned.

Other data were collected, digitized and georeferenced, if necessary, to analyse urban growth of the city. This entailed using the old cards to map the growth of Kinshasa over the long-term, population data and the relief and major roads.

In addition, to map the dynamics of the habitat of the Atlas of Kinshasa (Flouriot, 1975), the map "District Urban Leopoldville 1/60 000" presents the urban area in 1920. Plan Leopoldville (map 1/15 500 published by the bookseller Congo Leopoldville) gives the limit of the city in 1954. The map "Plan of Commons of Kinshasa and its Environs" to 1/20 000 published in 1959 by the Geographic Institute of Zaire is the drawing of municipal boundaries of the urbanized area in 1959. The map "City of Kinshasa-health zones" (Card 1/20 000 published in 1969 and revised in 1997 from the bottom of the base map of Kinshasa), provisional edition, published by the Geographical Institute of Zaire has the delineation of municipal boundaries of the urbanized area in 1969. All these documents are completely overwhelmed by the current situation (Delbart et al., 2002; Fox et al., 1997) and require updating.

The population data used suffers from both a paucity of quality and reliability in a country where the offices of the civil state are characterized by operating failure and where the general census of the population is not regularly organized. With the exception of the 1984 population numbers from the 1984 census, the others are mere projections of the National Institute of Statistics.

Coverage maps scale 1 / 10 000 by the Geographical Institute of the Belgian Congo (IGCB) dating from 1958 covering the city of Kinshasa have been scanned. The contours at a contour interval of 5 metres were digitized by students from MA1 geography at the university, corrected and interpolated by Mathieu De Maeyer (IGEAT / ULB) by the spline technique to produce a digital terrain model and derive the slope.

Some roads (in the north of the city and the far east, after the airport) were digitized from the SPOT panchromatic band (of 10 April 2000) and a plan of the city of Kinshasa (1 / 10 000) of March 1970 created by the Geographical Institute of the Congo. The roads in the west and south were measured and corrected by DGPS Pathfinder software. The railway was also digitized from the map of the city of Kinshasa. The roads of the southern part were digitized using only the SPOT panchromatic band of the 10 April 2000.

## 4. Methodology

Two approaches for change detection exist. "Image-image" comparison methods imply a radiometric normalization; this standardization is difficult to implement on data from different seasons and radiometric quality is also variable (Singh, 1986; Alphan, 2003; Coppin et al., 2004; Yuan et al., 2005). In addition, they do not identify the nature of change. Comparison methods compare the post-classification classifications of land produced independently at different dates (Gupta et al., 1985). The other group of methods is less sensitive to differences in season and they identify the nature of change but are susceptible to misclassification. To detect changes, classifications are compared in pairs. From this comparison, a map where the changes can be located and a change matrix that summarizes the amount and the nature of these changes are derived.

### 4.1 Geometric correction and cutting recovery images

To detect changes, it is essential that the SPOT images are properly stowed from the geometrical point of view.

This is why the latest panchromatic image has been corrected from an image of higher resolution. This is a panchromatic IKONOS image from 2002 of a resolution of 1 m corrected itself with control points measured in absolute mode with a Garmin GP60 GPS. Root mean square errors of 9.46 m on the hilly part and 4.14 m on the plain were obtained.

Then all the other images SPOT (panchromatic and multispectral mode) were corrected on the panchromatic SPOT image, corrected with a polynomial function of first order and the nearest neighbour method. All are projected onto the ellipsoid WGS 84 UTM coordinates, zone 33 south.

Geometric corrections lead to RMS errors smaller than the size of a pixel with 29 to 35 control points (Table 1), which is acceptable according to Moller-Jensen (1990) and is suitable for a detect changes study.

| Image | Cell (m) | Control points number | XRMS (cell) | XRMS (m) | YRMS (cell | YRMS (m) |
|---|---|---|---|---|---|---|
| Spot panchromatic 1995 | 10 | 30 | 0.44 | 4.4 | 0.57 | 5.7 |
| Spot multispectral 1995 | 20 | 34 | 0.49 | 9.8 | 0.54 | 10.8 |
| Spot multispectral 2005 | 20 | 29 | 0.53 | 10.6 | 0.44 | 8.8 |

Table 1. RMS errors after geometric correction

Not all SPOT images have the same spatial extension. In addition, their size being 60 km on each side, is wider than the extension of the city of Kinshasa. The images of 1995 and 2005 were cut to the same extension.

## 4.2 Land use classification

Given the uneven quality of SPOT images and the strong texture of the buildings, they were classified by a supervised method and object-oriented software using eCognition.

### 4.2.1 Legend

The legend distinguishes four categories: the built-up, the non built-up (vegetation and bare soil), water and clouds.

Some classes are difficult to discriminate using only spectral characteristics, especially so in countries in sub-Saharan Africa. The spectral confusions are numerous, for example, the fields are easily confused with the built-up. Production facilities and services, and the buildings for residential use in some places have the same spectral signature as the sand and burned areas. To overcome these problems, we have enriched the description of spectral regions of texture parameters (see 4.2.3).

### 4.2.2 Selection of training and validation areas

Training and validation areas were selected based on a visual interpretation of SPOT images supported by a consultation of Google Earth and the plan of the city of Kinshasa, and edited by Aquaterra Kin Art in 1997, ensuring changes due to differences between dates of these documents. 68 areas were selected in common areas of the SPOT images. To ensure an equivalent content of classes on each date, only areas unchanged between 1995 and 2005 were selected. The sample was divided into two, 34 areas for training and 34 for validation.

### 4.2.3 Choice of attributes

The attributes used in the classifications were chosen on the basis of visual interpretation. The regions are described in terms of spectral averages in each spectral band and the NDVI and the textural point of view, by the standard deviations on the green and red bands, and two textural parameters of Haralick (1973), such as homogeneity and entropy of the panchromatic band.

### 4.2.4 Segmentation and classification

eCognition was used to perform segmentation and classification. This software can simultaneously use a variety of data, panchromatic and multispectral images or vector data bases, and can create multiple levels of segmentation using a hierarchical approach.

The segmentation algorithm is the "multiresolution segmentation." According to the "Definiens Developer 7 User Guide" (2007), this algorithm merges the pixels into segments of image by minimizing the average heterogeneity and maximizing their respective homogeneities. It can do the same with image segments from a previous segment. The procedure iteratively merges the pixels or segments, as long as the maximum threshold of heterogeneity is not exceeded. Homogeneity is defined as a combination of spectral properties and form. The spectral homogeneity is based on the standard deviation of the distribution of the colour and consistency of form is based on the deviation from a compact or smooth (Cantou et al., 2006). The procedure can be influenced by the scale factor that limits the size of the resulting segments. The segmentation was performed on the image of 2005 spectral bands of green, red and near infrared respectively, giving them a weight of 2, 1 and 1. The scale parameter was chosen by trial and ,error and set at 20 with the shape parameter 0.1 (0.5 for compactness and 0.5 for smoothing).

The algorithm for supervised classification of the nearest neighbour was used. It ranks the regions according to their proximity to areas of statistical training.

### 4.2.5 Validation

The classifications are evaluated by comparing 34 areas of validation within the matrix of confusion. Indices are calculated to assess the quality of results (Richards, 1993):

- The overall accuracy,
- The overall Kappa,
- The Kappa class.

The overall accuracy is good (> 80%) obtained for the different classifications (Table 2). The Kappa coefficient is only acceptable for the classification of 1995 (85%) and 2005 (92%). The classification of 2000 has a poor Kappa (64%) caused by the fog that covers the southwest of the city. This result will therefore not be used subsequently.

| Years | Overall accuracy (%) | Kappa Coefficient (%) |
|-------|----------------------|-----------------------|
| 1995  | 93                   | 85                    |
| 2005  | 96                   | 92                    |

Table 2. Classification accuracy

Extensive field visits conducted in late January 2009 to the end March 2009 in the extension zones of Kinshasa to understand the factors of urbanization has revealed the existence of different confusions and omissions in the class "building". For example, here are some for the image of 1995 and 2005. They are located in Figure 1 and identified in Table 3.

| | Confusions errors in 1995 |
|--------|---------------------------|
| Zone 1 | Canoes to the east of Industrial Limete |
| Zone 2 | Island Mimosa |
| Zone 3 | Fields and sand pit in the southwest in the commune of Mont Ngafula |
| | **Confusions errors in 2005** |
| Zone 1 | Sand pit area behind the camp CETA and fields of vegetable crops |
| Zone 2 | Field burned to the east of the city |
| Zone 4 | Island Mimosa with large rocks and a mining company in building materials |
| Zone 5 | Sand bank to the west |
| | **Errors of omission in 2005** |
| Zone 3 | Residences in the area of the general staff of the Congolese armed forces (abundance of vegetation) |

Table 3. Confusion and omission errors for the class built-up in 1995 and 2005

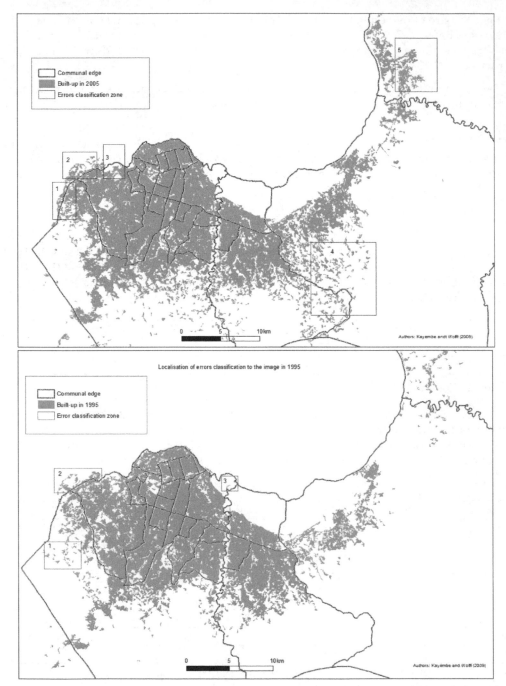

Fig. 1. (a and b): Location of misclassification (Source: Delbart and Wolff, 2002 for the map of municipal boundaries)

These errors will not be included in the analysis.

The results obtained at different dates are generalized by removing polygons classified as "built" with an area less than 1km² and the inclusion of less than 1 km² within the urban sprawl.

The superposition of classes "built" on two successive dates can map the evolution of the building when the matrix changes can be quantified.

## 5. Results

### 5.1 Urban growth

The location change was analysed using:

- Field visits conducted in-depth from the end of January 2009 through to March 2009 in the extension zones of Kinshasa to understand the factors of urbanization,
- The layout of the lines of roads and railways digitized,
- Maps and plans of the city of Kinshasa,
- Population data,
- The digital elevation model and slope map.

The map resulting from the comparison of land use classifications in 1995 and 2005 shows that the extensions of the city is concentrated in the southwest and northeast of Kinshasa (Figure 2).

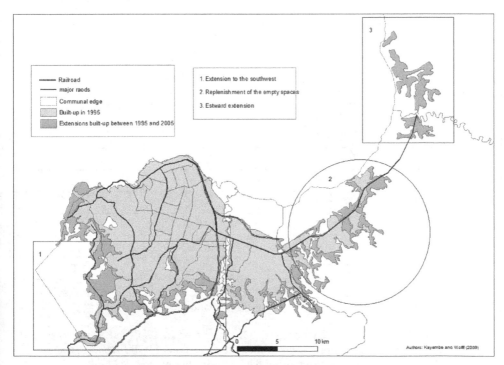

Fig. 2. Evolution of urban sprawl between 1995 and 2005 (Sources: Images Spot KJ 3 096-358 March 31, 1995 and KJ 4 096-358 July 01, 2005)

The spatial extension in the southwest took place mainly along the roads of Matadi and Lutendele (Zone 1). This process concerns the cities of Benseke, Kimbondo, Sans Fil and Matadi Mayo on the road to Matadi, and the cities of Lutendele, Kimbala, Zamba and Mazanza on the road to Lutendele. Cities such as Benseke and Kimbondo or Kimbala and Zamba have even joined in 2005.

To the east, there is a filling of interstitial spaces (Zone 2) and extension (Zone 3). Indeed, in neighbourhoods Mpasa I, II and III, Mikonga and the Badara camp, to the east of the River Ndjili, the blanks were filled. While in the far east, across the river Ndjili (Kinkole), the built-up was extended.

Urban growth can be explained by a population growth (5.1.1). Its spatial location can be explained by two main geographical factors beyond the simple distance to downtown, also an employment centre, the relief (5.1.2) and lines of communication (5.1.3).

## 5.1.1 Urban growth and population growth

Table 4 shows the evolution of the population of the extent of the city of Kinshasa and its density. In 45 years, the population rose from 400,000 to 7.5 million inhabitants in 2005, while the building area covered 6800 ha in 1960 against 43,400 in 2005. The population density tripled between 1960 and 2005 from about 60 inhabitants / ha in 1960 to 170 inhabitants / ha in 2005, on the whole, the city has expanded and become denser.

| Years | Population | Surface (ha) | Density (hab/ha) |
|-------|-----------|--------------|------------------|
| 1960 | 400000 | 6800 | 59 |
| 1967 | 901520 | 9470 | 95 |
| 1969 | 1051000 | 12903 | 81 |
| 1973 | 1323039 | 14600 | 91 |
| 1975 | 1679091 | 17992 | 93 |
| 1981 | 2567166 | 20160 | 127 |
| 1984 | 2653558 | 26000 | 102 |
| 1995 | 4719862 | 31007 | 152 |
| 2000 | 6000000 | 39518 | 151 |
| 2005 | 7500000 | 43414 | 173 |

Table 4. Evolution of the population, the extent of Kinshasa and its density (Sources: Lelo Nzuzi, 2008; Yebe Musieme, 2004; Delbart et al., 2002; Mbuila Matot, 2001)

When reporting on data on population and built-up areas in 1960, one can compare the growth in urban population. Figure 3 shows that the extension of buildings characterized by an index of 600 in 2005, while the population has an index of nearly 1900.

To compare growth rates, there is data on a semi-logarithmic graph (Figure 4).

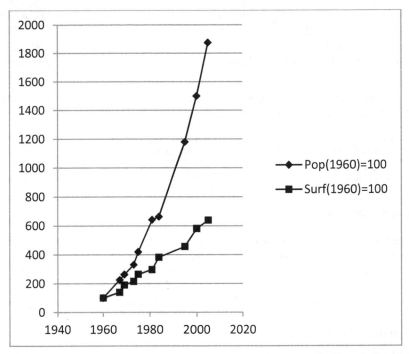

Fig. 3. Urban growth and population growth (Sources: Lelo Nzuzi, 2008; Yebe Musieme
Beni, 2004; Delbart et al., 2002; Mbuila Matot, 2001)

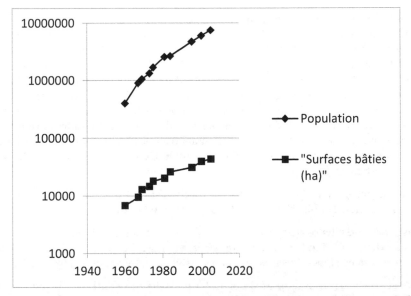

Fig. 4. Semi-logarithmic graph of urban growth and population growth (Sources: Lelo
Nzuzi, 2008; Yebe Musieme Beni, 2004; Delbart et al., 2002; Mbuila Matot, 2001)

Figure 4 shows that the average growth rate of the population is less than the extension of the city. The average growth rate of the population over the period 1960-2005 is 6.73%, while that of the built area is 4.21%. Applying this growth rate to the built area of 2005 to calculate the extension of the city in 2009, we do not get the 600 km² regularly cited, but only about 510 km².

By analysing the slopes, one can compare the growth rates, they both appear to decline in 1981 (Table 5). This result confirms the observations of Bruneau (1994).

| Growth rate | Population | Urban |
|---|---|---|
| 1960-1981 | 9.26% | 5.31% |
| 1981-2005 | 5.07% | 3.25% |

Table 5. Population growth rate

The first period covers the 20 years after independence (until 1981). It is characterized by a very high population growth and rapid expansion space. During the second period, the population growth rate slows sharply, although it remains high, from 9.26% between 1960 and 1981 to 7.5% between 1981 and 2005. The spatial extension grew at a slower pace and passes from 5.31% between 1960 and 1981 to 3.25% between 1981 and 2005.

### 5.1.2 Urban growth and relief

According to the observation of existing maps and plans (Figure 5), the extension to Kinkole phases that have characterized the spatial development of Kinshasa since its inception are confirmed.

Born in the west to the Bay of Ngaliema, the city had its first developments to the east with the birth of the ancient cities (Kinshasa, Barumbu, Lingwala, Kintambo) and Gombe (formerly Kalina) in the late 1920s. From then the city grew to the south with the birth of new cities (Kasavubu and Ngiri-Ngiri) between 1930 and 1940. It was during the 1950s that the city took over the management of the east with the merger of Kalina (Leo West) and Leo. Compared to the town of 1959, we find that the city is much more extensive in the south, southwest and east (beyond the communes of Kimbanseke and Masina). Comparing the growth of map altitudes and landforms, we observe that Kinshasa was first extended in the plain corresponding to the extension of the Malebo pool and until independence in 1960 (Figure 6), the colonial authorities strictly prohibited constructions on the hills in the absence of a particular development.

After independence, the city expanded to the southwest on the plateau to the east and the plains. For the period 1995-2005, the growth has continued in the same directions.

To the east, it extends the plain due to the narrow width of flat land in the east (Biyeye, 1997). Indeed, the extension behind the neighbourhoods Mpasa I, II, III and Mikonga did not take place because of steep slopes; this is how urban sprawl has moved beyond the city Kinkole.

To the west of the river Ndjili, areas of flat plains to the south are being built upon, urbanization covers steep slopes (Figure 7), but they are unfit for human settlement in the absence of appropriate management. Indeed, these areas of steep slopes are subject to

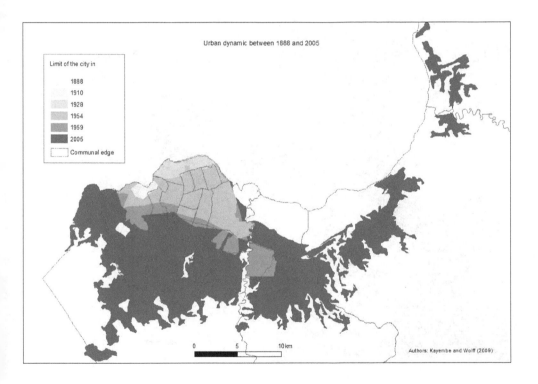

Fig. 5. Urban growth in Kinshasa from 1889 to 2005 (Source: historical maps collected by Johan Lagae, Department of Architecture and Urban Planning, Ghent University)

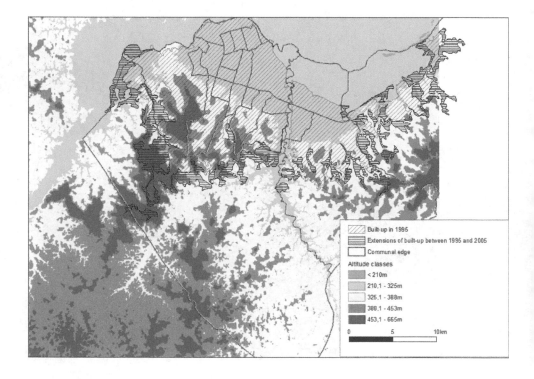

Fig. 6. Urbanization and altitude (Source: Mathieu De MaeyerIGEAT / ULB to the DTM, unpublished)

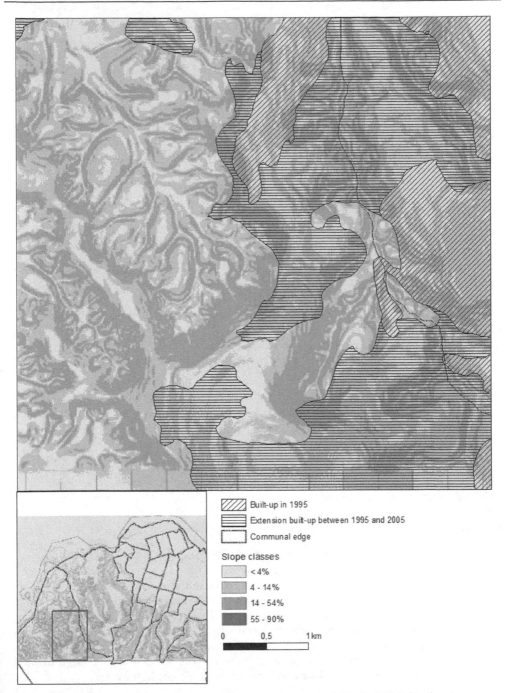

Fig. 7. Urbanization and slope (Source: Mathieu De Maeyer, IGEAT / ULB for the slope
map, unpublished)

significant risk of erosion as they are laid bare (Van Caillie, 1990, 1997). In addition, the plains downstream of these steep slopes are affected by floods because of silting. These areas contain steep slopes occupied by the poor. In the future, urbanization will continue to locate in areas of high slope, as is already happening in many places on the hillsides (Camping areas, Kindele, etc.).

### 5.1.3 Urban growth and major roads

Figure 2 shows that urban growth is more influenced by the roads along the railway. If before independence the railway played a role in the location of industrial areas, the urban railway had not developed, now it has not facilitated access to downtown as a centre of employment and therefore has not polarized urban growth. The roads in Matadi and Bandundu, as well as routes to the drop in Lukaya and to Lake Ma-Vallée, correspond to areas of urban growth today.

Despite the role of highways in urban growth, there are areas that develop latest far from downtown and away from these axes. Indeed, an urban extension area has been observed since the 1990s, south and east of the camp which houses the faculty of the University of Kinshasa. This is the area south Cogelos and neighbourhoods Tchad, Mandela and Department Plateau to the east. These areas develop in the absence of urban amenities. Indeed, they are connected to almost none of the service water supply of REGIDESO. The population is supplied fitted to the sources (Mayi ya Libanga, Mayi ya Niwa, Mayi ya Zamba) or the fountain. This is the case in the district of Mbiti. Where the water is high, people dig a well. The lack of urban amenities does not limit urban growth in Kinshasa. The bottom line for people is to have a home.

The quantitative analysis of urban growth compared to major communication axes, made in the GRASS software, shows that 47.5% of the growth took place at more than 1 km of main roads. Beyond this 1 kilometre threshold, the advantage of proximity to the main roads can be neglected and the neighbourhood effect becomes more important in the sense that people settled near existing neighbourhoods, but at a greater distance from the road.

This helps to highlight the fact that the major communication axes are not always, or are no longer, a major factor in urban growth.

## 6. Conclusions

Two high-resolution satellite images (SPOT) of 1995 and 2005 were used to map and quantify the urban growth in Kinshasa relatively quickly and with an acceptable reliability. The city spreads very quickly on its margins primarily to the east and southwest along the road to Matadi and Bandundu allowing access by public transport to the city centre which polarizes the bulk of urban employment. However, since the early 1990s, neighbourhoods are growing away from the city centre and transport routes (e.g. district Cogelo, Tchad, Mandela, Department, Plateau), yet they do not benefit from any urban convenience. The extension of the city after 1960 did not spare areas of steep slopes unfit for human settlement in the absence of a particular development. These areas are home to the poor.

By measuring the average growth rate of the population and the extension of the city over 45 years (from 1960 to 2005), it was found that it is 6.73% versus 4.21% for that of the built-up area. The average growth rate of built surface applied to the surface, built in 2005 to calculate the area built in 2009 revealed some significant errors with the figures regularly quoted.

In the future, a study could be carried out to understand the logic which pushes people to occupy the steeply sloping zones where the problem of gully erosion is acute.

## 7. References

Alphan H. (2003). Land use change and urbanization in Adana, Turkey, *Land Degradation and Development*, Vol.14, No.6, pp. 575–586, Online ISSN 1099-145X.

Biyeye Unzola E. (1997). Urbanisation de Kinshasa: la politique de la partie extrême orientale, Mémoire en Urbanisme et Aménagement, Université Libre de Bruxelles, 85 p.

Bruneau J-C. (1995). Crise et déclin de la croissance des villes au Zaïre – Une image Actualisée, *Revue Belge de Géographie*, *Belgeo*, 119e année, N° spécial offert au professeur H. Nicolaï, pp. 103-114, ISBN 1377-2368.

Cantou J. P., Maillet G., Flamanc D. & Buissart H. (2006). Preparing the use of pleiades images for mapping purposes: preliminary assessments at ign-france», Topographic Mapping from Space, ISPRS Workshop Commission I, WGI/5 and WG I/6, page 6.

Coppin P., Jonckheere I., Nackaerts K., Muys B. & Lambin E. (2004). Digital change detection methods in ecosystem monitoring: A review, *International Journal of Remote Sensing*, Vol.25, No.9, pp.1565–1596, ISSN 0143-1161.

Delbart, V. & Wolff, E. (2002). Extension urbaine et densité de la population à Kinshasa: contribution de la télédétection satellitaire, *Revue Belge de Géographie*, *BELGEO*, Vol.2, No.1, pp. 45-59, ISSN 1377-2368.

Flouriot, J., De Maximy, R. & Pain, M. (1975). *Atlas de Kinshasa*, Institut Géographique National du Zaïre.

Guptad, N. & Munshi, M.K. (1985).Urban change detection and land-use mapping of Delhi, *International Journal of Remote Sensing*, Vol.6, No.3-4, pp. 529-534, ISSN 0143-1161.

Haralick R.M., (1979). Statistical and structural approaches to texture, *Proceedings of the IEEE*, Vol. 67, No. 5, pp. 786-804, ISSN 0018-9219.

Lelo Nzuzi, F. (2008). *Kinshasa, ville et environnement*, L'Harmattan, 275p, ISBN 978-2-296-06080-7, Paris.

Moller-Jensen, L. (1990). Knowledge based classification of an urban area using texture an context information in LandSat-TM imagery. *Photogrammetric Engineering and Remote Sensing*, Vol.56, No.6, pp. 899-904, ISSN 0099-1112.

Pain, M. (1978). Kinshasa : Ecologie et organisation urbaines. Thèse de doctorat, Université de Toulouse le Mirail, 470 p+ Annexes.

Singh, A. (1986). Digital change detection techniques using remotely sensed data. *International Journal of Remote Sensing*, Vol.10, pp. 989-1003, ISSN 0143-1161.

Tshibangu, K.W.T., Engels, P. & Malaisse, F. (1997). Evolution du couvert végétal de la région de Kinshasa (1960-1987), *Geo-Eco-Trop*, Vol.21, No.1-4, pp.95-103, Liège.

Van Caillie X.D. (1990). Erodabilité des terrains sableux du Zaïre et contrôle de l'érosion, *Cahier ORSTOM*, série Pédologie, vol. 25, No.1-2, pp. 197-208.

Van Caillie X.D. (1997). La carte des pentes (1/20 000) de la région des collines à   Kinshasa. *Cahier ORSTOM*, réseau érosion, bulletin 17: Erosion en montagnes semi-arides et méditerranéennes, pp. 198-204.

Yebe Musieme B. (2004). L'impact des érosions sur l'habitat à Kisenso et les travaux de lutte anti-érosive par la population locale. Travail de fin de cycle, Unikin, Faculté des Sciences, Département des Sciences de la terre, Géographie, 63p, unpublished.

Yuan F., Sawaya K.E., Loeffelholz B.C. & Bauer M.E. (2005). Land cover classification and change analysis of the Twin Cities (Minnesota) Metropolitan Area by multitemporal Landsat remote sensing, *Remote Sensing of Environnement*, Vol.98, No.2 & 3, pp. 317-328, ISSN 0034-4257.

# 9

# Remote Sensing for Medical and Health Care Applications

Satoshi Suzuki and Takemi Matsui
*Kansai University, Tokyo Metropolitan University*
*Japan*

## 1. Introduction

Radar-based remote sensing techniques are typically employed to determine the velocities and positions of targets such as aircraft, ships, and land vehicles. In particular, X- and K-band microwave devices, including oscillators and antennas, have been used to measure the velocity of automobiles and other moving objects in recent years. Microwave devices that are compact, accurate, reliable, and inexpensive are currently commercially available. Over the past few years, there have been increasing attempts to apply such techniques to biomedical measurements. Although some studies have applied these devices to medicine and health care, such research is still in its infancy. This chapter focuses on the mechanisms of and the recent research trends in microwave remote sensing techniques that are used to detect minute vibrations on the body surface induced by heartbeat and respiration.

### 1.1 Background

The increasing proportion of elderly in the population represents an appreciable problem in developed countries due to social concerns such as increased medical and social welfare costs and a shortage of manpower. Such concerns are expected to worsen in the future. It is thus necessary to focus on preventing illnesses and to promote healthy lifestyles. Consequently, simple equipment that can be used to self-monitor medical conditions and to acquire related data is required for homes as well as medical facilities.

Vital signs are parameters of physiological functions that are used to express the physical condition. They are used by medical professionals for making initial diagnoses. There are four primary vital signs: heart rate, respiratory rate, body temperature, and blood pressure. Thermometers for home use are commercially available and are generally approved by medical bodies. In addition, heart rate and respiratory rate can be easily confirmed by visual and palpation methods. However, there is currently still not spread to home device capable of accurately measuring and recording vital sign data that can be used to make detailed diagnoses. Monitoring cardiac function can be used for diagnosing arrhythmia and mental stress (Akselrod et al., 1981, Singh et al., 1996, Carney et al., 2001). Recently, monitoring mental condition has attracted more attention than monitoring physiological parameters. And also obesity and aging are thought to contribute to the risk of developing sleep apnea syndrome (SAS). Airway obstruction due to fat deposition in the neck is one cause of SAS and it is related to reduced alertness during the daytime (Morriset al., 2008). A simple device that can monitor respiratory activity throughout the night is thus required. These examples show the necessity for monitoring of the vital signs in daily life. Moreover, these sensing technique are presently being studied in the research area on human–machine interfaces that can be applied anywhere (for example, in a car or at the workplace) (Sirevaag et al., 1993, Gould et al., 2009).

In addition, patients who have been exposed to toxic chemicals or infectious diseases are often treated in isolation chambers to prevent secondary exposure to health-care workers. In such cases, a doctor must often make a diagnosis without touching the patient, which is difficult as the vital signs are of primary importance for emergency medical treatment. With the exception of body temperature (which can be measured by infrared radiation), it is difficult to measure vital sign parameters without contact. Consequently, remote sensing of vital signs has attracted much attention.

In this way, several fields require remote sensing of vital signs and various remote sensing methods have been proposed. However, such methods should perform biomedical measurements described as non-invasive, non-restrictive, or non-contact means that can be used without the user being conscious of them. The use of physically attaching sensors should be minimized to reduce the burden on users.

## 1.2 Purpose and requirements of remote sensing in medicine and health care

Monitoring cardiac and respiratory parameters is useful for health-care management as users go about their everyday lives. However, such daily monitoring needs to overcome many problems. For example, users must have sufficient technical and medical competence to set electrodes to themselves and they must not feel physically restricted by the electrodes and leads. To overcome such problems, research is increasingly being conducted on developing non-invasive and non-restrictive sensing techniques for acquiring vital signs (Jacobs et al., 2004, Wang et al., 2006, Ciaccio et al., 2007). This kind of sensing technique aims to detect and measure vibrations on the body surface induced by cardiac and respiratory activity. In the case of respiratory activity, a person's abdomen expands and contracts during the breathing cycle and this movement can be detected by sensing techniques. Similarly, for cardiac activity, the body surface moves in response to the heartbeat in minute scales. Although the vibration is slight and its amplitude depends on the individual and the part of the body, it has been observed from all parts of the body with an average amplitude of about 0.1–0.2 mm by a high-resolution laser distance meter (Suzuki et al., 2011).

Some studies have measured heart rate by placing a pressure sensor (Jacobs et al., 2004) or polyvinylidene fluoride piezoelectric sensors (Wang et al., 2006) between the user and the mattress on which they sleep. This kind of measurement method measures responses to pressure changes. Other trials have used strain gauges to measure the heart rate (Ciaccio et al., 2007). The size of minute changes due to pressure changes on the body surface induced by the heartbeat and information relating to heartbeat and respiration were obtained. A similar procedure was employed in studies using air mattresses (Watanabe et al., 2005).

Such sensing techniques have the advantages of being inexpensive because of their simple structure and of enabling stable relatively stable data acquisition because they employ direct contact with the body. Some of these sensors are already commercially available. However, they suffer from one drawback: measurement is not possible when the sensor is separated from the body by moving their bodies. This raises the question: "Is a remote sensing method available?"

### 1.3 Biomedical measurement using microwaves

Radio-frequency sensing techniques were originally developed for military applications and they were used to determine the location and velocity of aircraft and ships. The same technology was then applied to search and rescue; for example, they have been used to locate survivors buried under earthquake rubble (Chen et al., 1986, 2000, Lin et al., 1992). Radar can remotely acquire information on the motion of targets. Additionally, depending on the frequency of the electromagnetic wave used, radar can penetrate barriers. These characteristics of radar have been employed to detect body motion of survivors under earthquake rubble. Such devices initially had very limited effectiveness because of their poor resolution by using low-frequency waves to penetrate rubble; they could only detect relatively large body motion (at best, the abdominal motion due to breathing). However, the permeability is not a problem for everyday applications since microwaves can readily penetrate materials such as clothing, bedding, and mattresses.

A cheap, small unit that is stable and can oscillate at high frequencies has recently been developed and ongoing development is being conducted to produce safer, more flexible devices. As a consequence, higher frequency electromagnetic waves were contributed to enhance the resolution of measurement. At the same time, the output power was reduced to reduce its effect on humans, allowing microwaves to be used in everyday applications. Gradual progress, therfore, has made it possible to detect even human heartbeat.

## 2. Theory and methods

Here, we describe a system that employs microwaves to remotely measure vital signs by detecting vibrations on the body surface induced by cardiac and respiratory activity. Vibrations induced by heartbeat are particularly small with amplitudes of about 0.1–0.2 mm on average. This section discusses approaches using continuous-wave (CW) Doppler radar and ultra-wideband (UWB) pulse radar, which are generally used for measuring vital signs, and their mechanisms.

### 2.1 Mechanisms of measurement

While frequency-modulated continuous wave (FMCW) radar is used to identify the exact location of a subject in some reports, UWB or CW Doppler radar are generally used for monitoring vital signs. (Saunders, 1990, Immoreev & Tao, 2008, Li & Lin, 2010)

In a UWB pulse radar, the transmitter sends very short electromagnetic pulses toward the target. A pulse duration of about 200–300 ps and a pulse repetition frequency in the range of 1–10 MHz are typically used for vital sign detection. When the transmitted pulse reaches the chest wall, some of the energy is reflected and captured by the receiver. The nominal round-trip travel time of the pulse is defined as $t = 2d/C$, where d is the nominal detection distance and $C$ is the speed of the electromagnetic wave. If a local replica of the transmitted pulse with a delay close to the nominal round-trip travel time correlates with the received echo, the output correlation function will have the same frequency as the physiological movement.

On the other hand, the CW Doppler radar mechanism is based on following (1);

$$T(t) = \cos\left[2\pi ft + \varphi(t)\right] \tag{1}$$

where an unmodulated signal $T(t)$ with a carrier frequency $f$ and a residual phase $\varphi(t)$, is transmitted toward a human body where it is phase-modulated by the physiological movement $x(t)$. The reflected signal $R(t)$ detected by the radar receiver is given by following (2);

$$R(t) \approx \cos\left[2\pi ft - \frac{4\pi d_0}{\lambda} - \frac{4\pi x(t)}{\lambda} + \varphi\left(t - \frac{2d_0}{c}\right)\right] \tag{2}$$

where $4\pi d_0/\lambda$ is a constant phase shift due to the nominal detection distance $d_0$ and the $\varphi(t - 2d_0/c)$ is phase noise. Using the same transmitted signal $T(t)$ as the local oscillator signal, the radar receiver down-converts the received signal $R(t)$ to the baseband signal $B(t)$ as following (3);

$$B(t) \approx \cos\left[\frac{4\pi d_0}{\lambda} + \frac{4\pi x(t)}{\lambda} + \theta_0 + \Delta\varphi\right] \tag{3}$$

where $\Delta\varphi$ is determined by the nominal detection distance and the oscillator phase noise.

Since the delay corresponds to the signal round-trip travel time, the detection range of a UWB radar can be varied by controlling the delay between the two inputs of the correlation function block. This makes it possible to eliminate interference caused by reflection from other objects (clutter) and multipath reflection. However, one disadvantage of UWB radar is that the delay needs to be recalibrated when the detection distance is changed; this increases the system complexity and cost. Furthermore, since the correlation function is nonlinear, it is not simple to recover the original movement pattern, even though frequency information can be easily obtained. On the other hand, CW Doppler radar has a low power consumption and a simple radio architecture. These characteristics make it suitable for home-based systems. Moreover, proper adjustment of the radio front-end architecture of a CW radar can cancel clutter (Li & Lin, 2008a, 2008b). In addition, single-input multi-output and multi-input multi-output techniques can be easily implemented with CW radar, enabling the movements of multiple targets to be detected (Boric-Lubecke et al., 2005, Zhou et al., 2006).

## 2.2 Carrier frequency and output power

The carrier frequency and output power employed must be safe for use on people. Carrier frequencies ranging from hundreds of megahertz to millimeter wave frequencies have been tested for remote vital sign detection using a variety of physiological movements. The carrier frequency should be carefully selected to ensure suitable sensitivities and characteristic response for vital sign measurement. Some studies used extremely high-frequency waves (228 GHz (Petkie et al., 2009)), which have shorter wavelengths and are more sensitive to small displacements. Moreover, a 228 GHz frequency prototype has been extended to perform heart rate and respiration measurements at a distance of 50 m. However, such high frequency waves are not realistic for monitoring vital signs in everyday applications. In many cases, a carrier frequency that does not require a license is often chosen. However, carrier frequencies that do not require a license vary from country to country and some frequency bands are allocated to amateur radio stations. For example, the laws regulating radio frequency use in Japan allow band frequencies 10.525 and 24.15 GHz to be used for detecting moving objects. These devices are marketed as sensors for measuring the speeds of vehicles. Although there are limitations on how they are used (e.g., limited to indoor use), these frequency bands can be used by low-power radio stations without a license provided the output is less than 10 mW. They have been increasing studies on frequency bands for vital sign monitoring. Regarding safety, different countries have different guidelines regarding radio-frequency electromagnetic fields.

The World Health Organization (WHO) and the Scientific Committee on Emerging and Newly Identified Health Risks (SCENIHR) define exposure as *the subjection of a person to electric, magnetic, or electromagnetic fields or to contact currents other than those originating from physiological processes in the body and other natural phenomena* (WHO, 2003, SCENIHR, 2006). The intended frequency band of electromagnetic field intensity differ slightly in different guidelines. For example, the International Commission on Non-Ionizing Radiation Protection (ICNIRP) guidelines (ICNIRP, 1998) specify the frequency band from 300 Hz to 300 GHz, SCENIHR specifies 100 Hz to 300 GHz (SCENIHR, 2006), and the IEEE Standard is from 3 kHz to 300 GHz (IEEE Standard Committee, 1998). Each country employs different methods for determining their criteria. Consequently, it is important consider device development and intended usage.

The carrier frequency of medical applications of ultrasound is a low frequency of about 3 to 10 MHz. Such applications acquire information by penetrating the human body. In comparison, monitoring using microwave frequencies in the range 10.525 to 24.15 GHz is considered to be less invasive and safer. It is not easy to make simple comparisons, but wireless local area networks (WLANs) use 2.4 GHz radio waves and microwaves in the range 10.525 and 24.15 GHz are considered to be safer. Moreover, safety can be further increased by using a lower power than a WLAN.

It seems appropriate to use high frequencies for sensing to ensure a high resolution while considering invasiveness. However, high frequencies are not necessarily ideal for actual applications as increased sensitivity results in increased susceptibility to artifacts; the target motion induced by heartbeat on the body surface is much smaller than the artifacts generated by general movement of the body and arm. Furthermore, people being monitored move freely as they conduct everyday activities, which makes artifacts a significant problem.

## 3. Examples of applications

While remote sensing is not currently used for medical and health care applications in everyday life, several studies have been conducted. This section discusses the following typical examples of remote sensing in medical and health care settings: (1) monitoring daily health and mental stress by estimating changes in the autonomic nervous system (ANS); (2) welfare and health care for the elderly; and (3) medical diagnosis such as screening of patients with infections.

## 3.1 Measurement of ANS

To determine stress levels when driving or operating equipment, a 24 GHz compact microwave radar was used to perform remote measurements of the heart rate variability (HRV) under autonomic activation induced by a stressful sound and foods (Suzuki et al., 2008, Gotoh et al., 2009). Changes in the ANS (sympathovagal balance) can be monitored by measuring the HRV. This variability is used as an index of mental stress.

For examples, when an animal is attacked, its sympathetic nervous system spontaneously prepares for fight or flight by elevating the heart rate and increasing the blood pressure and body temperature. In contrast, when it is in a relaxed state, the parasympathetic nervous system is activated, and the opposite phenomena occur. Such processes are constantly in equilibrium in daily life (Akselrod et al., 1981, Derrick,1988).

Sympathovagal balance can be monitored by measuring variations in the heartbeat interval, and the degree of mental stress can be determined. This index is used for diagnosing psychological disorders and as well as the condition of circulatory organs. Moreover, many approaches for measuring the HRV are used in psychology (Vincent et al., 1996), occupational health (Miyake, 2001, Princi et al., 2005), and ergonomics research (Sirevaag et al., 1993, Gould et al., 2009). An estimation technique has been medically established. Thus, if electrocardiograms (ECG) can be replaced by a remote sensing method that allows simple and accurate monitoring in everyday life, mental stress can be assessed at any time and in any place.

24GHz
Microwave
radar antenna

Fig. 1. Prototype chair equipped with a non-contact microwave radar system to monitor mental stress in workers (Suzuki et al., 2008).

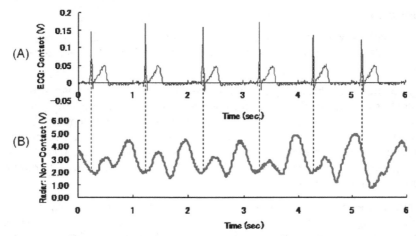

Fig. 2. Sample data of a compact microwave radar output (A) showing a cyclic oscillation that corresponds to cardiac oscillations measured by ECG (B) (Suzuki et al., 2008).

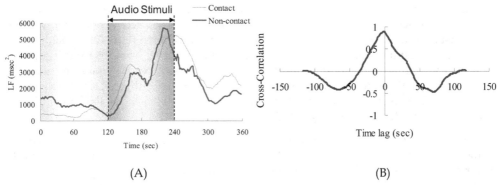

(A)                                                        (B)

Fig. 3. Example of remote sensing in health care. (A) In both non-contact and contact (ECG) measurements, the LF of a subject (reflecting sympathetic activation) exhibits a peak during audio stimulation. (B) Cross-correlation of the LF between non-contact and contact measurements of the same subject (Suzuki et al., 2008).

The prototype device in this example consists of a compact 24 GHz microwave Doppler radar (8 × 5 × 3 cm) attached to the back of a chair. The low-frequency/high-frequency (LF/HF) components of the HRV (which reflects the sympathovagal balance) of eight volunteers were determined by a prototype system using the maximum entropy method (MEM). The change in the LF and LF/HF components measured by a remote sensor was consistent with that measured by a contact electrocardiography sensor. It is very difficult to extract R-R intervals of heartbeats sufficiently accurately to calculate the HRV; this is thought to be because inaccuracies were introduced due to the microwave radar being susceptible to noise. However, measuring HRV by adopting MEM, which enables HRV to be estimated rapidly, stably, and accurately, was demonstrated to be successful for the first time.

## 3.2 Elderly care

The next examples were monitoring of the elderly, infants and also animals using a microwave radar, especially focused respiratory monitoring (Li et al., 2009, Suzuki et al.,2009).

There is a real need to reduce the physical and mental work load of care workers and also to immediately detect sudden changes in the condition of a bedridden elderly person, both at home and in hospitals, because nighttime activity of persons with dementia increases the risk of injury and disrupts the sleep patterns of caregivers. Moreover, care of the elderly will increasingly be undertaken by family members at home as the population continues to age. Therefore, devices for long-term monitoring of the elderly that do not interfere with activities of daily life are required.

Various approaches have been proposed in recent years. Rowe (Rowe et al., 2009) proposed a night monitoring system that alerts caregivers when care recipients leave their beds and that tracks them as they move about the house during the night. In addition, another study compared the effects of a self-care and medication compliance device, which was linked to a web-based monitoring system, with the effects of conventional care alone on compliance with recommended self-care behavior (Artinian et al., 2003). Demiris (Demiris et al., 2009) investigated elderly behavior using a video and image processing system while carefully addressing privacy concerns. Another study proposed a non-restrictive, non-invasive vital sign measurement system for measuring heartbeat and respiration to monitor health status at home or in hospitals and nursing facilities (Tanaka et al., 2002). Systems for nursing care should meet the following requirements: (1) monitoring of activity and vital signs must not be a burden for the elderly; (2) accurately monitor changes in physical condition of the patient; and (3) when the physical condition of a patient rapidly changes, the system should promptly notify a doctor or nursing caregiver.

A prototype system using microwaves has been developed and has been applied for monitoring elderly in a nursing home. Although still a trial, the system can effectively perform real-time monitoring and it can acquire SAS respiratory data.

In addition, a baby also monitor using same technology has recently been demonstrated (Li et al., 2009). The baby monitor integrates a low-power Doppler radar that can detect minute movements induced by breathing. If no movement is detected within 20 s, an alarm goes off.

Using same type of sensing devices, respiratory activity of a hibernating black bear has been monitored at a Zoo (Suzuki et al., 2009). Ueno Zoological Gardens in Tokyo made plans to assist a Japanese black bear to enter hibernation, because the bear showed extremely slow movements in winter. Moreover, the staff wanted visitors to understand this instinctive behavior of bears as occurs in the wild. The bear's condition during hibernation must be carefully observed to avoid the risk of long-term fasting. Therefore, to observe the physiological condition of the bear during hibernation, about 3 months, a microwave radar system was set up in the hibernation booth. As a result, the respiratory rate decreased while the bear was entering hibernation, and became extremely low at approximately 2 bpm and showed almost no change. Additionally, a trend similar to a circadian rhythm in the changes of respiratory rate was observed.

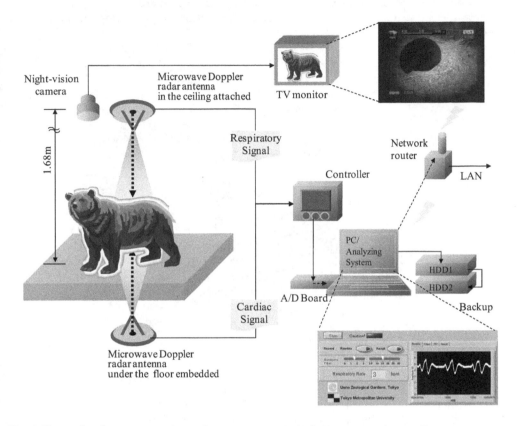

Fig. 4. Example of remote sensing using non-contact vital sign sensors to monitor a hibernating black bear at Ueno Zoo, Tokyo (Suzuki et al., 2009).

### 3.3 Screening and medical examination of influenza patients

The advantage of monitoring without direct touch and without removing clothing is useful and appealing in the medical field as it is not burdensome to patients and it can reduce the risk of secondary infection. There is also growing interest in the fields of health, life science, and engineering.

In another trial, a new screening system was developed to conduct rapid screening (<5 s) of passengers who may have infectious diseases such as severe acute respiratory syndrome (SARS) or pandemic influenza at quarantine stations. This system enables medical inspection by measuring heart and respiratory rates, as well as body temperature by infrared thermography (Matsui et al., 2009, 2010). A similar system is already being tested at the quarantine station at Naha Airport for domestic flights and at Narita International Airport in Tokyo. The results demonstrate the efficacy of the concept and the system. The system even detected a patient with influenza whose fever was reduced by antifebrile medication. This finding is important as it demonstrates that the parameters for monitoring cardiac and respiratory activity by microwaves are effective. In the future, remote sensing using microwaves is anticipated to attract more interest in the medical field.

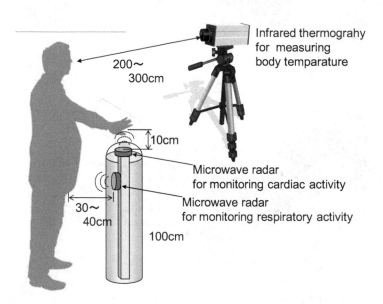

Fig. 5. Example of remote sensing for medical diagnosis; non-contact monitoring for screening systems at airport quarantine stations (Matsui et al., 2009).

## 4. Topics for future investigation

Remote sensing of biomedical parameters has been described by focusing on monitoring using microwave radar and examples of medical and health care applications have been presented. Microwave-based remote sensing offers the following advantages: movement of an object can be detected from a distance, it means "remotely", and microwaves can pass through many materials (notable exceptions are metals and water). Thus, such systems will enable health care workers to measure motion of a body surface through clothing from a distance.

The most serious obstacle to practical applications of this technology is the need to reduce the effects of noise and artifacts. Vibrations on the body surface caused by heartbeat are quite small (with amplitudes of about 0.1–0.2 mm), whereas motions of arms and the abdomen are much larger. Additionally, movement of the person being monitored while conducting every day activities and data from other body motion will generate noise. The effect of multipass reflection should also be considered. One study performed simultaneous measurements of the vital signs of two people, but it is difficult to apply this technology to real-world applications.

To use this technique in medical fields and daily life, aspects such as non-contact measurement and the non–removal of clothing need to be addressed. Data obtained by remote sensing must be more reliable and accurate than those measured by conventional electrode-based methods. Furthermore, a new index for diagnosing disease is required. Most studies of remote sensing using microwaves only measure the heart rate. However, the heart rate alone has limited diagnostic value; the heart rate only provides information about the tachycardia or bradycardia. It is hoped that such a system will be able to measure parameters for diagnosing cardiovascular disease.

In conclusion, research on remote sensing using microwaves is still in its infancy, but it appears promising not only for medical and health care, but also for other fields such as ergonomics research.

## 5. References

Akselrod, S. et al. (1981) Power spectrum analysis of heart rate fluctuation: a quantitative probe of beat-to-beat cardiovascular control. Science, 213(4504), pp.220-222.

Artinian, N.T., Harden, J.K., Kronenberg, M.W., Vander Wal, J.S., Daher, E., Stephens, Q., Bazzi, R.I. (2003) Pilot study of a Web-based compliance monitoring device for patients with congestive heart failure. Heart Lung, 32(4), pp.226-233.

Boric-Lubecke, O., Lubecke, V., Host-Madsen, A., Samardzija, D., Cheung, K. (2005) Doppler radar sensing of multiple subjects in single and multiple antenna systems. 7th International Conf. on Telecom. In Modern Satellite, Cable and Broadcasting Services, 1, pp. 7-11.

Carney, R.M., Blumenthal, J.A., Stein, P.K., Watkins, L., Catellier, D., Berkman, L.F., Czajkowski, S.M., O'Connor, C., Stone, P.H., Freedland, K.E. (2001) Depression, heart rate variability, and acute myocardial infarction. Circulation, 104, pp.2024-2028.

Chen, K.M., Misra, D., Wang, H., Chuang, H.R., Postow, E. (1986) An X-band microwave life-detection system. IEEE Trans Biomed Eng, 33, pp.697-702.

Chen, K.M., Huang, Y., Zhang, J. (2000) Microwave Life-Detection Systems for Searching Human Subjects Under Earthquake Rubble or Behind Barrier. IEEE Trans Biomed Eng, 27, pp.105-113.

Ciaccio, E.J., Hiatt, M., Hegyi, T., Drzewiecki, G.M. (2007) Measurement and monitoring of electrocardiogram belt tension in premature infants for assessment of respiratory function. Biomed Eng Online 6, pp.1-11

Demiris, G., Oliver, D.P., Giger, J., Skubic, M., Rantz, M. (2009) Older adults' privacy considerations for vision based recognition methods of eldercare applications. Technol Health Care, 17(1), pp.41-48.

Derrick, W.L. (1988) Dimensions of operator workload. Human Factors, 30, pp.95-110.

Gotoh, S., Suzuki, S., Kagawa, M., Badarch, Z., Matsui, T. (2009) Non-contact determination of parasympathetic activation induced by a full stomach using microwave radar, Medical & Biological Engineering & Computing, 47, pp.1017–1019.

Gould, K.S., Roed, B.K., Saus, E.R., Koefoed, V.F., Bridger, R.S., Moen, B.E. (2009) Effects of navigation method on workload and performance in simulated high-speed ship navigation. Appl Ergon 40, pp.103-114.

ICNIRP Guidelines (1998) For Limiting Exposure to Time–Varying Electric, Magnetic and electromagnetic Fields (UP TO 300 GHZ), Health Physics, 74(4), pp.494–522.

IEEE Standards Coordinating Committee 28 on Non-Ionizing Radiation Hazards (1998) IEEE Standard for Safety Levels With Respect to Human Exposure to Radio Frequency Electromagnetic Fields, 3 kHz to 300 GHz. C95.1-1991 and C95.1a-1998.

Immoreev, I., Tao, T.-H. (2008) UWB radar for patient monitoring. IEEE Aerospace and Electronic Systems Magazine, 23(11), pp.11-18.

International EMF Project of World Health Organization (2003) Framework for Developing EMF Standards.

Jacobs, J., Embree, P., Glei, M., Christensen, S., Sullivan, P. (2004) Characterization of a novel heart and respiratory rate sensor. Conf Proc IEEE Eng Med Biol Soc, 3, pp.2223-2226.

Li, C., Lin, J. (2008) Complex Signal Demodulation and Random Body Movement Cancellation Techniques for Non-contact Vital Sign Detection. IEEE MTT-S International Microwave Symposium Digest, pp.567-570.

Li, C., Lin, J. (2008) Random Body Movement Cancellation in Doppler Radar Vital Sign Detection. IEEE Transactions on Microwave Theory and Techniques, 56(12), pp.3143-3152.

Li, C., Cummings, J., Lam, J., Graves, E., Wu, W. (2009) Radar remote monitoring of vital signs. IEEE Microwave Magazine, 10(1), pp. 47-56.

Li, C., Lin, J. (2010) Recent Advances in Doppler Radar Sensors for Pervasive Healthcare Monitoring. Proceedings of Asia-Pacific Microwave Conference 2010, pp.283-290.

Lin, J. C. (1992) Microwave sensing of physiological movement and volume change: A review. Bioelectromagnetics, 13, pp.557-565.

Matsui, T., Hakozaki, Y., Suzuki, S., Usui, T., Kato, T., Hasegawa, K., Sugiyama, Y., Sugamata, M., Abe, S. (2010) A novel screening method for influenza patients using a newly developed non-contact screening system, Journal of Infection, 60(4), pp.271-277.

Matsui, T., Suzuki, S., Ujikawa, K., Usui, T., Gotoh, S., Sugamata, M., Abe, S. (2009) Development of a non-contact screening system for rapid medical inspection at a quarantine depot using a laser Doppler blood-flow meter, microwave radar, and infrared thermography, Journal of Medical Engineering & Technology, 33(6), pp.481-487.

Miyake, S. (2001) Multivariate workload evaluation combining physiological and subjective measures. Int J Psychophysiol 40, pp.233-238

Morris, L.G., Kleinberger, A., Lee, K.C., Liberatore, L.A., Burschtin, O. (2008) Rapid risk stratification for obstructive sleep apnea, based on snoring severity and body mass index.Otolaryngol Head Neck Surg, 139(5), pp.615-618.

Obeid, D., Issa, G., Sadek, S., Zaharia, G., El Zein, G. (2008) Low power microwave systems for heartbeat rate detection at 2.4, 5.8, 10 and 16 GHz. First International

Symposium on Applied Sciences on Biomedical and Communication Technologies, pp.1-5.

Petkie, D.T., Benton, C., Bryan, E. (2009) Millimeter wave radar for remote measurement of vital signs. IEEE Radar Conference, pp.1-3.

Princi, T., Parco, S., Accardo, A., Radillo, O., DeSeta, F., Guaschino, S. (2005) Parametric evaluation of heart rate variability during the menstrual cycle in young women. Biomed Sci Instrum 41, pp.340-345.

Rowe, M.A., Kelly, A., Horne, C., Lane, S., Campbell, J., Lehman, B., Phipps, C., Keller, M., Benito, A.P. (2009) Reducing dangerous nighttime events in persons with dementia by using a nighttime monitoring system. Alzheimers Dement, 5(5), pp.419-426.

Saunders, W.K. (1990) CW and FM Radar. in Radar Handbook, 2nd ed. M.I.Skolnik) San Francisco, MaGraw-Hill, Inc. pp.14.1-14.45

Scientfic Committee on Emerging and Newly Identified Health Risk (SCENIHR) (2006) Possible effects of Electromagnetic Fields (EMF) on Human Health.

Singh, N., Mironov, D., Armstrong, P.W., Ross, A.M., Langer, A. (1996) Heart rate variability assessment early after acute myocardial infarction. Pathophysiological and prognostic correlates. Circulation, 93, pp. 1388-1395.

Sirevaag, E.J., Kramer, A.F., Wickens, C.D., Reisweber, M., Strayer, D.L., Grenell, J.F. (1993) Assessment of pilot performance and mental workload in rotary wing aircraft. Ergonomics 36, pp.1121-1140.

Suzuki, S., Matsui, T., Imuta, H., Uenoyama, M. (2008) A novel autonomic activation measurement method for stress monitoring: non-contact measurement of heart rate variability using a compact microwave radar. Medical & Biological Engineering & Computing, 46(7), pp. 709-714.

Suzuki, S., Matsui, T., Kawahara, H., Gotoh, S. (2009) Development of a Noncontact and Long-Term Respiration Monitoring System Using Microwave Radar for Hibernating Black Bear. Zoo Biology, 28(3), pp.259–270.

Suzuki, S., Matsui, T., Kawahara, H., Ichiki, H., Shimizu, J., Kondo, Y., Yura, H., Gotoh, S., Takase, B., Ishihara, M. (2009) A non-contact vital sign monitoring system for ambulances using dual-frequency microwave radars, Medical & Biological Engineering & Computing, 47(1), pp.101-105.

Suzuki, S., Matsui, T., Sugawara, K., Asao, T., Kotani, K. (2011) An approach to remote monitoring of heart rate variability (HRV) using microwave radar during a calculation task, Journal of Physiological Anthropology, 30(6), pp.241-249.

Tanaka, S., Matsumoto, Y., Wakimoto, K. (2002) Unconstrained and non-invasive measurement of heart-beat and respiration periods using a phonocardiographic sensor. Med Biol Eng Comput, 40(2), pp.246-252.

Vincent, A., Craik, F.I., Furedy, J.J. (1996) Relations among memory performance, mental workload and cardiovascular responses. Int J Psychophysiol 23, pp.181-98.

Wang, F., Tanaka, M., Chonan, S. (2006) Development of a wearable mental stress evaluation system using PVDF film sensor. J Adv Sci, 18, pp.170-173.

Watanabe, K., Watanabe, T., Watanabe, H., Ando, H., Ishikawm, T., Kobayashi, K. (2005) Noninvasive measurement of heartbeat, respiration, snoring and body movements of a subject in bed via a pneumatic method.IEEE Trans Biomed Eng, 52(12), pp.2100-2107.

Zhou, Q., Liu, J., Host-Madsen, A., Boric-Lubecke, O., Lubecke, V. (2006) Detection of multiple heartbeats using Doppler radar," IEEE ICASSP 2006 Proceedings, 2, pp.1160-1163.

# 10

# Object-Based Image Analysis of VHR Satellite Imagery for Population Estimation in Informal Settlement Kibera-Nairobi, Kenya

Tatjana Veljanovski[1,2], Urša Kanjir[1], Peter Pehani[1,2],
Krištof Oštir[1,2] and Primož Kovačič[3]
[1]*Scientific Research Centre of the Slovenian Academy of Sciences and Arts*
[2]*Space-SI – Centre of Excellence for Space Science and Technologies*
[3]*Map Kibera Trust*
[1,2]*Slovenia*
[3]*Kenya*

## 1. Introduction

Cities in Africa and developing countries in general are having a difficult time coping with the influx of people arriving every day. Informal settlements are growing, and governments are struggling to provide even the most fundamental services to their urban populations.

Kibera (edge region within the Nairobi) is the biggest informal settlement in Kenya, and one of the biggest in Africa. The population estimates vary between 170,000 and 1 million and are highly debatable. What is certain is that the area is large (roughly 2.5 km²), host at least hundreds of thousands people, is informal and self-organized, stricken by poverty, disease, population increase, environmental degradation, corruption, lack of security and - often overlooked but extremely important – lack of information which all contribute to lack of basic services such as access to safe water, sanitation, health care and formal education.

In Africa, but also in other continents, urban growth has reached alarming figures. Informal settlements formation has been associated with the rapid growth of urban population caused by rural immigration, triggered by difficult livelihood, civil wars and internal disturbances. The result of this very rapid and unplanned urban growth is that 30% to 60% of residents of most large cities in developing countries live in informal settlements (UNHSP, 2005). Nowadays, informal residential environments (slums) are an important component reflecting fast urban expansion in poor living conditions.

Densely populated urban areas in developing countries often lack any kind of data that would enable the monitoring systems. Monitoring systems joining spatial (location) and social data can be used for the monitoring, planning and management purposes. New methods of monitoring are required to generate adequate data to help link the location and socioeconomic data in urban systems to local policies and controlling actions. In the past, rapid urban growth was quite difficult to manage and regulate when processes were in progress. Available census data barely accounts for the reality, as in most cases, they are based on figures extrapolated from old census, carried out in the 1970s or, if recent, they are obtained with poor accuracy, as informal settlements are difficult to survey (Sartori et al., 2002). More can now be done at least to monitor the extent and consequences of rapid urban growth. Where accurate maps of informal settlements and relevant census data completely lack, answers can be found using independent survey, derived from satellite or aerial technologies. Usage of satellite imagery nowadays enables rather quick answers to questions such as: where informal settlements are, what was the dynamics of their growth, how many people potentially live there, what basic services inhabitants need. Among the main issues to be addressed in informal settlements are the needs for potable water, waste evacuation, energy, education and health care facilities, and crime control. It is believed these actions can be planned based on quality mapping of the phenomena.

The spatial resolution of space-borne remote sensing has improved to such extent that their products are comparable with the ones provided by aerial photography. Satellite images taken with very high resolution (VHR) sensors, i.e. resolution around and below 1 m, enable skilled user to identify and extract buildings, trees, narrow paths and other objects of comparable size. A side effect of higher resolution is larger quantity of data which require more storage capacities and processing costs. Detection of informal residential settlements from satellite imagery is especially challenging task due to the microstructure, merged/overlapping rooftops and irregular shapes of buildings in slum-like areas. High spatial resolution is essential to facilitate extraction of individual buildings that are characterized by small, densely packed shanties and other structures. Informal settlement Kibera is composed of varying sizes of houses, where roofs can be a combination of many different materials, and mainly unpaved road and path network. Typically this can produce a spectral response on satellite imagery that is difficult to interpret and makes it difficult for traditional classification strategies to differentiate across object class type.

Various approaches enable to extract data from imagery in urban environments. Simultaneously with expansion of VHR satellite systems an object-based image analysis (OBIA) was developed to answer new technological opportunities. OBIA approach works in similar way as human brain perceives nature/environment, namely (high detailed) image is segmented into homogeneous regions called segments or "image objects" (Benz et al., 2004), which are then classified into meaningful classes, following the specific context of the study.

### 1.1 Objectives of the research

Objective of the work perform was to help Map Kibera Trust initiative with satellite data processing. Studies on Kibera informal settlement had two aims: first, to derive detailed land use/cover map that can further supply population estimation, and second, to analyse the potential of VHR imagery for detecting changes and settlement growth in recent past.

Since object-based classification of VHR satellite data has been argued as the most appropriate method to obtain information from urban remote sensing applications, this approach was used to derive accurate land cover map. The study involved GeoEye and QuickBird satellite images acquired between 2006 and 2009. Object-based approach was used to determine detailed urban structure in informal settlements area. Urban expansion was analyzed through comparison of images taken on different dates, using contextual multi-level pixel based approach. The results of object-based analysis based on morphology attributes were further explored to estimate the potential population. There is a big discrepancy among estimations on Kibera population, thus different density parameters were tested to approach the potential population scenario.

The first, introductory chapter sets the informal residential settlement issue in the wider context of the remote sensing possibilities framework, highlighting the methodology of the study. Chapter 2 gives an overview of research and applications of informal residential environments monitoring. Chapter 3 reviews existing conditions in Kibera, Nairobi's informal residential settlement, bringing into perspective the historical development of the slum, and its current characteristics. Chapter 4 consists of a set of specific procedures performed at two spatial extents, to attain both aims of the study. Entire Kibera settlement was being reviewed, to map the general state and dynamics of housing (change detection) between years 2006 and 2009. Raila village was studied in detail using object-based analysis to derive precise map of the village land cover/use to derive population estimation models in a given situation. Chapter 5 collects the results of mapping and population estimations. Chapter 6 discusses the data and analyses involved in managing monitoring aspects of the slums. The last chapter concludes the study with some suggestions for future work.

## 2. Informal residential environments monitoring

Although there is a strong need to obtain spatial information about informal settlements in order to increase living conditions for its residents and regarding the fact that remote sensing images offer a well suited data source, studies on informal settlements with VHR data are not frequent. Nevertheless, in Hoffman (2001), first results of detecting informal settlements from IKONOS data in Cape Town showed the principle feasibilities using object-oriented approach. The results were promising but seemed to be very dependent on the data. Later on Hoffman et al. (2006) showed that several adaptations were necessary to OBIA algorithm improvement when applying their extraction methods to the QuickBird scene. Automatic image analysis procedures for a rapid and reliable identification of refugee tents from IKONOS imagery over the Lukole refugee camp in Tanzania was made by Giada et al. (2002). Sliuzas and Kuffer (2008) analyzed the spatial heterogeneity of informal settlements using selected high resolution remote sensing based spatial indicators such as roof coverage densities and a lack of proper road network characterized by the irregular layout of settlements. Cooperation between KeyObs, UNOSAT, OCHA and Metria resulted in digitalization of VHR GeoEye satellite image of Afgooye corridor (Somalia) from 2009, where all temporary shelters were identified (UNHCR, 2010). Different methods to detect and monitor spatial behaviour of informal settlements were presented also by Lemma et al. (2005), Radnaabazar et al. (2004), Kuffer (2003), Sartori et al. (2002), Dare & Fraser (2001) and Mason et al. (1998).

## 3. Study area description

Kibera is a division of Nairobi area, Kenya, within Langata constituency. Located southwest of the city centre of Nairobi, Kibera encompasses an area of 2.5 km², accounting for less than percent of Nairobi's total area while containing more than 25% of its population. It is the largest informal settlement in Nairobi, and the second largest urban slum in Africa, with population number varying with the season. The settlement is divided into a number of villages, including Kianda, Soweto West, Raila, Gatwekera, Kisumu Ndogo, Lindi, Laini Saba, Siranga, Kamdi Muru, Makina, Mashimoni and Soweto East (Fig. 1).

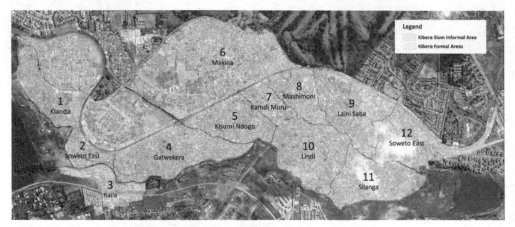

Fig. 1. Kibera settlement is divided into three formal and 12 informal villages.

### 3.1 General background of Kibera, Nairobi

Kibera emerged in 1912 when the British East African army, known as the King's African Rifles, granted temporary rights to a group of 300 former soldiers from the Nubian community, who had served in the army, to settle on a small piece of land near Nairobi's city centre. Temporary structures were put in place but as the Nubian soldiers grew older and became unable to continue their military service, they began to set up more permanent residence on the land (A history of Kibera, 2011). Turbulent years after the independence combined with socioeconomic factors brought a dramatic increase in the population of Kibera's residents.

Today Kibera consists of 15 villages out of which just 3 are formal and thus connected to the city's utility grids (water, sewage, electricity, waste collection etc.), however the rest (12) are informal and "disconnected" from the rest of the city. Apart from lacking basic services and adequate infrastructure it is also affected by population growth, the illegal construction of infrastructure, and the increasing degradation of the environment. Unclear land-tenure arrangements in informal settlements discourage investments in proper infrastructure and repair; structures are often owned and rented by people, who mostly do not have any rights to the land on which the structures stand. This leads to the lack of legal security of tenure for most of the residents.

Because of this lack of the legal security of tenure and neglectfulness from the city and the government there's little initiative from the residents to improve their living conditions. That is why most of the structures in Kibera are temporary, wooden, mud houses covered with corrugated iron sheets (Fig. 2) and most of the service providers are self-organized groups or cartels which drive up the prices of service delivery – in some cases residents pay 10 times as much as those in the rest of the city.

Object-Based Image Analysis of VHR Satellite Imagery for Population Estimation in Informal Settlement
Kibera-Nairobi, Kenya

175

Fig. 2. View over Kibera informal settlement (Photo: Primož Kovačič).

All these reasons lead to, as one resident of Kibera put it, "survival tactics". These "survival tactics" engulf communities, the provincial administration and the government, leading them into a vicious cycle of under the table dealings, vandalism, lack of engagement, threats, and price controls with no clear perspective or solutions.

## 3.2 Map Kibera Project and Map Kibera Trust

Kibera is likely one of the most photographed, researched, and well-known slums in the world but the complete and mapped information was not shared and not easily (if at all) accessible. Before October 2009, Kibera did not even appear on any of the online maps. Map Kibera Project (MKP) was first initiated in response to the lack of available data. The initiative wanted to produce reliable data and maps showing the actual physical and socio-demographic features of the Kibera informal settlement, making them publicly available through a digital geo-referenced data base (MKP, 2011).

Map Kibera Project trained 13 youth from the slum in GPS system and basic GIS techniques to map points of interest in their community: clinics, schools, water sources, toilets, street lights, hot spots, businesses and other landmarks. The youth uploaded the data themselves to OpenStreetMap (www.openstreetmap.org), a volunteer-built map of the world.

The Map Kibera Trust (MKT) offers organizational support for the mapping work, as well as other youth driven programmes such as video production and other new media tools (blog, twitter, SMS platforms). The mission of the MKT is to contribute to a culture where digital story-telling, open data and geographic information lead to greater influence and representation for marginalized communities in Kenya. MKT has since grown into a platform specializing in community-driven data for informal settlements and on community based development.

American Association for the Advancement of Science supports the operation of MKT and other NGO activities and has donated several satellite images of the area. MKT activities include various Kibera specific phenomena mappings (www.mappingnobigdeal.com), though the assessment of potential of VHR satellite imagery for mapping purposes presents one of the recent examinations of their use for Kibera community.

## 3.3 Available VHR satellite data

Six VHR satellite images were available for our research (Table 1): one GeoEye image and five QuickBird images. Satellite images were partly (pre)processed. This means images were roughly georeferenced and corrected for sensor radiometry, also pan-sharpened, and provided as a stack of three visible bands only.

| Date | Sensor | Bands used | Spatial resolution | Cloud coverage | Analysis performed |
|------|--------|-----------|--------------------|----------------|--------------------|
| 2006-03-27 | QuickBird | R-G-B | 0.6 (pansharpened) | minor | Change detection |
| 2006-07-31 | QuickBird | R-G-B | 0.6 (pansharpened) | free | |
| 2007-01-22 | QuickBird | R-G-B | 0.6 (pansharpened) | present | |
| 2008-01-07 | QuickBird | R-G-B | 0.6 (pansharpened) | free | |
| 2008-08-10 | QuickBird | R-G-B | 0.6 (pansharpened) | present | Change detection |
| 2009-07-25 | GeoEye | R-G-B | 0.5 (pansharpened) | free | Land use/cover Change detection |

Table 1. List of available satellite images and their main characteristics.

Besides different inherent spatial resolution the main differences among GeoEye and QuickBird images were sensor viewing angles, causing higher objects roof prints and shadows to have different positions among images. As Kibera informal settlement lies in a hilly terrain, the positional accuracy fit of geographical entities among images was not reached because much of distortion comes from the terrain as well. For the study no digital elevation model was available, thus ortorectification was not possible. However, GPS field walks tracks were available for the main roads and path-network in the area.

## 4. Methods

Study of Kibera informal settlement has two main aims: to derive detailed land use/land cover map that can supply population estimation, and to analyse the settlement growth and changes between 2006 and 2009.

Extracting data of urban land use structure from remote sensing imagery require methods that are able to provide appropriate level of details observed. Object based classification has been successfully implemented to obtain land cover information from urban VHR remote sensing applications. Thus, this approach was selected in the land use/cover classification of Kibera settlement with main aim to delimitate well residential objects from open areas, and potentially to obtain informal settlement structure in the microstructure level (distinguish individual houses). In addition to determination of detailed urban structures we were also interested in locating the step-wise expansion of informal residential areas, which was analyzed through comparison of images taken in different time using pixel-based multi-level image differencing approach.

Object based classification of the Kibera informal settlement was performed on GeoEye image since its characteristics (close to nadir viewing angle, good spatial resolution, and fine contrast) were most promising to obtain adequate details on object recognition within the informal settlement area. Rooftops are covered with different materials, ranging from new to rusty sheets, bricks and other materials, each of them having specific reflectance characteristics (spectral representation) on satellite image (Fig. 3, Fig. 6a). For population estimation study we need to differentiate well rooftops, unpaved roads and non-build land and therefore discriminate residential areas from open soils, respectively. Object based segmentation automatically delimits satellite image into homogeneous elements (segments), where close correspondence to the real (geographical) objects on the Earth's surface is expected. Usage of thus obtained image elements (segments) has a number of benefits, one of them is ability to incorporate spatial and contextual information such as size, shape, texture and topological relationships (Blaschke et al., 2004; Benz et al., 2004) in contextual classification. In the stage of classification all these segments are classified according to their attributes into most appropriate classes (representing various geographical objects under study consideration), while obtaining detailed classification of urban area land cover/use.

QB 2006-03-27                          QB 2008-08-10                          GE 2009-07-25

Fig. 3. Examples of rooftops, rooftops renovations and buildings constructions on VHR satellite imagery from three different dates.

With object-based analysis on rooftops morphology attributes we expected to improve the assessment of the potential population in slum areas. Since no complete and relevant field survey (official census) was recently performed, different density parameters were tested to approach the potential population and compared to other available population assessments.

## 4.1 Data pre-processing and preparation

Data preprocessing is important procedure in remote sensing technology. It meets issues that have to be carefully understood and solved before any data analysis process starts. In order to be able to compare satellite images taken for the same scene at different acquisition dates they have to be co-registered and radiometrically adjusted. Recent automatic registration algorithms can accomplish the task well when similar acquisition geometry among sensor systems is provided. Global geometric transformations are mostly appropriate for positional corrections in such cases. However, this was not the case with the imagery obtained for Kibera study (see section 3.3).

Obtained images were rectified but not precisely aligned one to another. Due to agitated terrain in Kibera and lack of any digital elevation model, semi-automated rigorous procedures of co-registration could not be applied. The non-linear rubber-sheeting method was the only possibility to obtain mutually aligned images. This procedure is effective, but very time consuming, since it demands manual selection of hundreds of control points for each image.

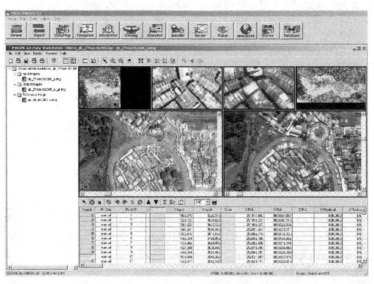

Fig. 4. Selecting the control points for rubber-sheeting method for image geo-referencing (AutoSync module of ERDAS Imagine).

GeoEye image taken on 25th of July 2009 was selected as the reference image considering its highest spatial resolution, good matching with GPS path-network tracks and the fact it is most recent. Then QuickBird images selected for analysis were manually registered to the reference, based on cca. 1,400 manually selected control points per image and using a piecewise transformation based on triangles formed from the tie points (Fig. 4). Resampling was nearest neighbour. An average RMSE is not reported as this is local approximation technique. Geo-corrected images were evaluated through detailed visual control.

Geometrically matched GeoEye and QuickBird images were then used for OBIA. Finally three images were selected for change detection due to their best results in geo-correction phase: GE2009-07-25, QB2008-08-10 and QB2006-03-27. For change detection analysis images were resampled to uniform 1 m resolution, to be prepared for radiometric standardisation.

After geometric adjustments there are still differences amongst the spectral properties of satellite images (spectral bands from the same or different images are not adjusted to each other). Hence, before pixel-based image comparisons (image differencing) the radiometric standardisation is needed. Most standardisation procedures derive from adjustments of invariant objects or from the least squares method (linear regression). The problem with the first method is that invariant areas should be verified with field measurements. Furthermore, the generally recognised invariant objects, such as deserts or light sandy beaches, do not come into play, since they cannot be found in Kibera area. The principle of the second approach is that it tries to globally adjust the given (to-be-adjusted) image or a chosen area with the reference image or a subset through a statistical approach. We applied linear regression for the relative adjustment of spectral bands between the images. Relative radiometric normalisation was done through local adjustments of QuickBird images onto GeoEye reference image, for Kibera settlement with 30 m buffer subset only.

## 4.2 Land cover classification

Since GeoEye image of the whole Kibera informal settlement contains lots of information and the analysis of the total area would be too demanding in terms of computer processing, we divided image of Kibera into 12 smaller parts (according to 12 informal villages). The complete process of segmentation and classification was applied systematically for each informal village separately, with same parameter settings at each phase. This was possible due to relatively homogenous landscape over Kibera settlement. Thus we obtained 12 regional classification outputs, which were then merged in the final stage. To avoid erroneous classification on the edges of the splitted images, we applied 30 meter buffer when masking the village fragments out from the whole GeoEye image (Fig. 5).

Fig. 5. A buffer of 30 meters around the village border.

Object based classification consists of two stages. Image is first segmented into a set of segments (regions) that are considered to be homogeneous in terms of one or more spectral or spatial properties. Then follows classification where each segment is classified into belonging object class.

Supervised segmentation within software used (ENVI EX, Feature Extraction module) is defined by two segmentation parameters that influence an average size of segments: segmentation and merging. Setting different values for these two parameters causes change of size of segments, allowing for an image to be segmented at many different scales, so both parameter values influence classification results. Since structure of the Earh's surface is similar throughout the whole settlement, same general segmentation parameters were used for each of the 12 villages. Visual example of segment structures are shown on Fig.6.

While classifying Raila village we adapted segmentation parameters to best extract shapes of individual buildings inside informal settlement. Since segmentation parameters were adequate for one particular land use only (i.e. buildings), others were expected to be under- or over-segmented. There is no single "optimal" scale for analysis of remote sensing images, rather there are many optimal scales that are specific to the image-objects that exist within a scale (Hay et al., 2003) and this is why using a multi-scale approach may often be preferable (Johnson and Xie, 2011). All spectral bands were used and given equal weight for image segmentation and all available attributes were calculated for all segments.

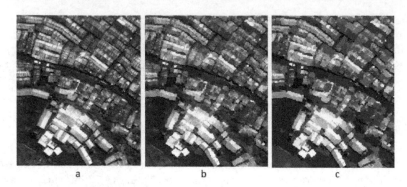

Fig. 6. Original GeoEye image (a), objects segmented (b), objects merged (c) (Feature Extraction module of ENVI EX).

The objects extracted during the segmentation were then classified using Support Vector Machine (SVM) classification algorithm in an object-oriented framework along with training sets, selected by experienced user. Nine land cover classes were used all together (Table 2).

| Land use / cover classes | Description |
|---|---|
| Buildings_blue | Residential houses with blue spectral reflectance on image. |
| Buildings_light | Residential houses with white or bright spectral reflectance. |
| Buildings_brown | Residential houses with brown or dark spectral reflectance. |
| Buildings_red | Residential houses with red spectral reflectance. |
| Roads | Traffic connection between villages, usually unpaved. |
| Shadows | Shadowed areas around high objects (high vegetation and buildings). |
| Soil | Areas of unvegetated soil, mudded ground. |
| Vegetation 1 | Green vegetated areas, low vegetation (grass). |
| Vegetation 2 | Green vegetated areas, high vegetation (trees). |

Table 2. Land cover classes anticipated with object-based classification.

These nine urban land use/cover classes included four types of residential housing. Sub-classes were chosen because of their different spectral signature inside the same land cover class (e.g. instead of selecting only class "buildings" we selected four subclasses "buildings_blue", "buildings_light", „buildings_brown" and "buildings_red", Fig. 7). This way we obtained better results than we would have using one general class only. More detailed classification of residential housing was made only in Raila village.

Classification results were obtained as a raster image and a vector file. Vectors were exported to a single layer and later processed for the need of post-classification in ESRI ArcMap software (all polygons smaller than 2 m² were merged with neighbouring larger polygons).

Fig. 7. Selecting training samples for supervised classification (Feature Extraction module of ENVI EX).

## 4.3 Change detection

Satellite data offers unique utility for monitoring and quantifying land cover change over time. Consequently, change detection has become a significant part of the remote sensing research over the last decades. The goal of remote sensing change detection is to detect the geographic location of changes, identify their type (if possible) and quantify their amount. If long term imagery time series are handled, trends can be recognised. A large number of change detection methods have been evolved and they differ in their refinement, robustness and complexity (Hall and Hay, 2003). Nowadays a three level systematisation system is proposed that differentiates change detection methods by introducing the notion of pixel, feature and object level image processing (Deer, 1998). In general change detection techniques can be grouped into two major types (Jianya et al., 2008; Coppin et al., 2004; Lu et al. 2004; Singh 1989): image differencing techniques and post-classification comparison techniques. The main difference between the two types is that image differencing methods can identify the location and the magnitude of change but can not identify the type of land use or surface changes taken place in the area. Post-classification techniques can identify the location and provide the change character. Recent advances in change detection mostly involve high resolution data and consequently object-oriented and/or multi-scale approaches, with a range of techniques to approach contextual modelling (Lang et al., 2006; Blaschke et al., 2008; Addink & Van Coillie, 2010).

### 4.3.1 Change detection of Kibera informal settlement

The use of VHR satellite image time series may provide a reliable approach to detect dense urban growth in detail (Hofmann et al., 2006). A generically applicable and rapid operational land cover mapping of these settlements has generally proven difficult (Netzband & Rahman, 2010). Object-based classification and land use mapping of Kibera settlement from GeoEye image (section 4.2 and 5.2) highlighted some typical problems for object delineation in slum-like areas that can be corrected only with a lot of manual work. Main difficulties are associated with informal area outer-homogenity but inner-heterogenity due to the microstructure of urban agglomeration. Object-based classification is thus very demanding in terms of methodology adaptation to informal residential areas specifics, especially when accounting for their direct relation to representation on different satellite data sources. Thus within the limited framework of this case study pixel-based approach to identify outline of urban growth was preferred. The procedure was implemented on radiometrically adjusted time series (section 4.1). GeoEye 2009-07-25 and QuickBird 2006-03-27 images were compared for the changes over whole Kibera, and GeoEye 2009-07-25, QucikBird 2008-08-10 and 2006-03-27 images were analysed for the observation of sequential urban growth of Raila village.

The simple thresholding of difference images is a well-known method that leads to the delimitation of changes and no-changes. The advantage of this method is that it can be fully automatic. However main disadvantage is that it heavily depends on consistency of datasets. Regardless of the carefully performed data preparations certain unwanted effects remains, which may drastically burden the imagery comparisons. This data variability behaves as a detected change and may well be enclosed within identified pattern of changes, although not all of the identified differences belong to real changes (false, non-intrinsic changes). Such false effects result in over-estimation of change pattern and can cause the quantitative evaluation fail. Since this data noise originates from the pre-processing algorithms as well as the natural and technological conditions during data acquisition, it can not be completely removed with data radiometric corrections (Veljanovski & Oštir, 2011).

To overcome this drawback a contextual multi-level change detection approach was applied that can efficiently treat most of the unwanted differences and suppress sensor related noise (Veljanovski, 2008). Taking into account the neighbourhood and change information by joining two spatial scales (Fig. 8), approach reduces amount of small size false differences.

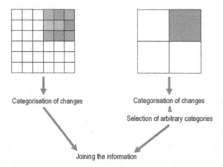

Fig. 8. Change detection approach takes into account the neighbourhood and change information by joining different spatial scales.

The model is based on focal information logic that gives averaged change information in a slightly reduced spatial scale – within the specified neighbourhood. It is based on the fact that in a larger geographic area (e.g. a 3 x 3 pixel window or 3 m spatial resolution for resampled VHR data) the information of the changes will tend to level abrupt change information if a small spatial scale change is present, and will show the averaged difference if the majority of pixels in the observed window are subjected to change. Computing the piecewise change information between two time-successive data sets provides valuable information regarding the location and numeric change value derived from contextual information within the specified neighbourhood.

The procedure was implemented as follows. First, spectral information for a slightly coarser scale (i.e. 3 m spatial resolution) was computed for images or areas of interest. This may be accomplished with a specified neighbourhood mean value annotation. Second, change differentiation between images is performed on a coarser resolution scale and change magnitude categorisation is applied (see below for categorisation classes and their definition). Third, upper positive and negative changes are reclassified so that the mask of important changes based on the neighbourhood context characteristics is prepared. Fourth, change differentiation is calculated on the original data scale (i.e. GeoEye and QuickBird data in 1 m spatial resolution), then categorisation is applied, and finally a mask (or a mask with a buffer) of an arbitrary specified magnitude of changes (obtained in the previous step) is overlaid in order to restrict merely the contextually supported changes.

Normally, if there is no substantial unwanted effects (noise) due to meteorological or sensor influences in images, changes obtained from the difference image (image differentiation) are distributed normally and symmetrically, with the average at 0. Abrupt changes (objects or land cover transformations) can then be defined by thresholding the distribution tails, giving the pattern of positive and negative changes in reflectance (spectral) space. We have calculated change magnitude (transition class category) intervals for every 0.5 standard deviation. Then the criterion of 2.5 standard deviations for negative changes and 2.0 for positive changes was applied to enclose the majority of detected transformations in both real and spectral world situations. Result of such categorisation of image difference is a pattern of abrupt changes (locations of appearance, disappearance of objects), with no association to change characterisation (type of change, from-to).

Each output from the above automatic threshold procedure is finally refined to an arbitrary degree, depending on case study objective. For entire Kibera informal settlement changes were obtained from comparison of GeoEye 2009-07-25 and QuickBird 2006-03-27 images. Change patches smaller than 5 m² were eliminated for the purpose of this example, mainly to reduce the impact of change artefacts belonging to small patches of rooftop renovations (Fig. 3). False changes identified due to differences in viewing angle (location of buildings and trees shadows, buildings boundaries due to different original resolution of imagery) were also removed using the results obtained from object based classification (state of land use in 2009, section 4.2). Where shadow or vegetation object class were present, change pattern was corrected in the given context. Described and implemented step-wise change pattern refinement is shown in Fig. 9, for a subset of Raila village and northern neighbourhood. In other words, with simple generalisation we could control many aspects of the change pattern for study's specific aims.

Results were visually examined and evaluation of change pattern characteristics (over- and under-estimations) was done throughout the area using complementary comparison of satellite images involved.

### 4.3.2 Raila village change detection

Raila village is known to have undergone extensive development during recent years (Kibera Wikipedia, 2011; MKP, 2011). Thus temporally detailed examination was performed observing its urbanisation. Change detection procedure described in section 4.3.1 was in addition implemented for Raila village only, but for two time sequences: 2006-2008 and 2008-2009. QB 2006-03-27, 2008-08-10 and GE 2009-07-25 images were used for this example.

2006-2009 3-band difference image for Kibera subset (Raila village).    Change pattern over 2006-2009 3-band difference image.

Change pattern after automatic multi-level threshold procedure.    Change pattern after eliminating false changes due to shadows differences.    Change pattern after eliminating small patches due to rooftop renovations.

Fig. 9. Overview of change pattern intermediate results through implemented processing steps.

## 4.4 Population estimation

Population statistics gives very important information for understanding of modern society. Demographic research is one of the main research directions of social science for a better understanding of the interactions between population growth and social, economic and environmental conditions. The collection of population data depends mainly on the census, which is labour-intensive, time-consuming and demands high financial resources. The 2009 Kenya Population and Housing Census reported Kibera's population to be 170,070 (Karanja, 2010). This report was far from the belief of that time that Kibera slum was of the biggest informal urban settlements in the world. Several actors had provided and published over the years increasing estimations of the size of its population, most of them stating that it was the largest slum in Africa with the population exceeding 1 million. According to Davis (2006), a well known expert on urban slums, Kibera had a population of about 800,000 people. International Housing Coalition (IHC, 2007) talked about more than half a million people. UN-Habitat (2004) had released several estimations ranging between 350,000 and 1 million people. These statistics mainly come from analyses of aerial images of the area. IRIN (2006) estimated a population density of 2000 residents per hectare. In 2008 an independent team of researchers began a door-by-door survey named Map Kibera Project (MKP, 2011). A trained team of locals, after having developed an ad-hoc surveying methodology, has so far gathered census data of over 15,000 people and completed the mapping of 5000 structures, services (public toilets, schools), and infrastructures (drainage system, water and electricity supply) in the Kianda village. Considering data collected for Kianda village, the population of the whole Kibera slum can be estimated between 235,000 and 270,000 people.

Nonetheless, no estimation so far guessed by the MKP, or the UN, or the Government of Kenya or by other actors can be taken for granted and does not represent the real dimension of the population of Kibera. In general, no estimation can be proved nor refuted until an exhaustive census will be taken throughout the whole slum (Kibera Wikipedia, 2011).

Because population is not directly related to land cover surface reflectance, population estimation is still a challenging task based purely on remote sensing spectral signatures. Although population is not directly measurable on the remote sensing images this technology may provide good approximation of population estimation by measurement of visible variables, e.g. the number of residential buildings and/or the area of build-up zone (Zhang, 2003). There exist many studies using different approaches on remote sensing data for population estimation. Studies date from the early 1970s onward, where air photos were utilized for manual counts of dwelling units. There are three most used methods of population estimation by remote sensing: residence count method, area (density) method and regression model method (Zhang, 2003). Residence count method was mostly done in first period of studies on this topic on the western urban environment (Horton, 1974, Barrett & Curtis, 1986). Area density method was used by H.H. Wang (1990), F.Z. Wang (1990), P. Sautton (1998), Langford et al. (1994), Z.J. Lin (2001) and others. Regression model method is currently also often used (Galeon, 2010, Dengsheng et al., 2006, Zhang, 2003). With each type of method some ancillary field survey data are needed.

Considering the above situation and the fact that for Kibera we lack other potential socio-geographic data (elsewhere applied to predict population with regression technique), we decided to assess the population on residential land cover class information obtained from object-based classification with density per area method solely. For each village a total area of buildings was calculated and different occupation scenarios (i.e. persons/living area) were tested to observe the range of possible population fluctuation.

## 5. Results

With object-based (contextual) classification performed on GeoEye image with Feature Extraction module of ENVI EX, it was possible to obtain accurate land cover map and following this, total residential area of Kibera slum and its divisions (villages) with very high accuracy. From this data, those related to build-up areas, were used for population estimation. With multi-level contextual change detection implemented in Erdas Imagine, it was possible to obtain representative change pattern reflecting where in informal settlement intensive urbanisation processes have taken place. Results are presented in the following order: land cover mapping, change pattern identification and population estimation, for Kibera informal settlement (section 5.1) and Raila village (section 5.2), respectively.

### 5.1 Kibera

### 5.1.1 Kibera land cover map 2009

Object based classification and post-classification on GeoEye 2009-07-25 image was performed for each village in Kibera informal settlement separately. Finally individual results were joined in a land cover map of Kibera informal settlement (Fig. 10).

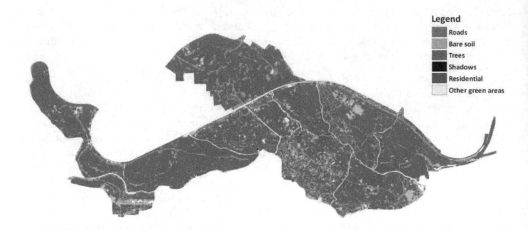

Fig. 10. Merged final classification results of GeoEye image where all 6 selected land cover classes are shown for all 12 Kibera villages (ESRI ArcGIS).

Vectorized classes of entire Kibera were merged together in order to be able to calculate area of total land use/land cover. As it is seen from Table 3, residential areas cover 2/3 (66%) of the whole Kibera area and are prevailing when compared to other land uses. This can be well confirmed from the visual examination of the satellite images of the discussed area.

Accuracy assessment was done by comparing results of supervised classification with manually digitalized objects. Comparison was done on the area 200 x 300 m in the village

| Land use [m²] / village | Residential | Trees | Green areas | Soils (roads, bare ground) | Shadows | Total |
|---|---|---|---|---|---|---|
| Kianda | 117,710 | 8,144 | 9,511 | 16,775 | 18,005 | 170,145 |
| Soweto West | 49,620 | 4,089 | 6,455 | 15,001 | 6,177 | 81,343 |
| Raila | 45,657 | 8,835 | 17,066 | 30,886 | 6,139 | 108,583 |
| Gatwekera | 234,609 | 17,634 | 11,718 | 19,682 | 23,878 | 307,520 |
| Kisumi Ndogo | 111,472 | 6,979 | 6,944 | 10,891 | 28,432 | 164,717 |
| Makina | 303,599 | 53,844 | 11,992 | 40,474 | 34,801 | 444,710 |
| Kamdi Muru | 51,248 | 1,829 | 7,541 | 12,123 | 8,670 | 81,412 |
| Mashimoni | 96,287 | 5,520 | 2,018 | 7,628 | 16,900 | 128,355 |
| Laini Saba | 181,211 | 17,400 | 13,141 | 45,545 | 20,490 | 277,786 |
| Lindi | 159,112 | 2,5826 | 11,885 | 48,215 | 26,630 | 271,668 |
| Silanga | 150,058 | 21,569 | 19,975 | 34,833 | 17,298 | 243,733 |
| Soweto East | 174,200 | 6,238 | 15,618 | 29,302 | 22,823 | 248,181 |
| SUM [m²] | 1.674,784 | 177,906 | 133,863 | 311,354 | 230,245 | 2.528,152 |
| SUM (%) | 66.25 | 7.04 | 5.29 | 12.32 | 9.11 | 100,00 |

Table 3. Area of different land use types for 12 informal villages and the total sum in Kibera.

Lindi. Only residential segments were estimated. Since with ENVI EX classification the outline of individual residential objects could not be extracted, we compared only the total sums of areas classified as *residential*. Results are shown in Table 4. The best result (error of 3%) was obtained when choosing parameter values for segmentation/merge: 85/85.

All (semi)automatic classification methods display some errors, but as an approximate solution object based classification on VHR data yielded very good results upon selection of proper segmentation parameter values. Different shapes and colours in informal settlements determine a complex urban formation, which is difficult to differentiate from other land cover types, especially from bare soils and unpaved streets. With the object-based classification dwelling zones were found with high accuracy all over the image, in spite of their spectral similarities with streets and other urban features, especially bare soils.

| | Area [m²] | Coverage [%] |
|---|---|---|
| Testing rectangle (200m x 300m) | 60,000 | |
| Manual digitalization | 37,913 | |
| **Supervised classification ENVI EX (segmentation 85, merge 85)** | **36,663** | **97** |
| Supervised classification ENVI EX (segmentation 85, merge 65) | 42,172 | 111 |
| Supervised classification ENVI EX (segmentation 85, merge 45) | 41,174 | 109 |

Table 4. Comparison of results of ENVI EX supervised classification using different segmentation parameters with manual digitalization. The total areas were compared for residential classes only.

### 5.1.2 Kibera change detection

The multi-level contextual change detection method we used in this case study was first developed for monitoring various greater and intensive processes on the Earth's surface with middle resolution imagery (i.e. Landsat, SPOT). The result on Kibera proved same contextual logic is efficient also when implemented on very high spatial resolution imagery.

Change detection applied to Kibera informal settlements aimed to obtain an outline of the distribution and extent of major urbanisation processes. Comparison was made for QB 2006-03-27 and GE 2009-07-25 images. Method has proven to be suitable for monitoring changes related to various processes (buildings construction, buildings collapse or disappearance, rooftop renovation, increase/decrease in vegetation) and/or of the coincident description of their trends (see section 5.2.2 for identified informal settlement growth in Raila village area). Although quantitative assessment of change occurrences was not performed due to lack of independent reference data, a detailed visual control was made through comparison of before and after images and, nevertheless, several conclusions can be derived.

Identified pattern of changes (Fig. 11) clearly draw attention to spots where urbanisation between years 2006 and 2009 was most intensive: Kianda north-east, Raila southern border, Mashimoni and Laini Saba northern border and Soweto East eastern tail. According to the density of change pattern elements, in addition to the edges of Kibera informal settlement, several changes occurred in the eastern part of Kibera. These are mainly due to larger new buildings constructions or complete rooftops renovations. Additional socio-economic data or bigger events information (like flooding) would be welcomed to associate the rate of more intensive rooftops renovations at some of Kibera villages and compared to the other parts of the settlements.

With external data, for example land cover for a reference point in time, a fairly reliable "from-to" change statistics could be extracted. However in rather homogenous land use areas like Kibera informal settlement is, where most of the land use is residential (section 5.1.1) and where urbanisation direction limits are well known due to formal residential settlement boundaries, such information would not additionally characterize the change pattern. On the example of Raila village change detection in section 5.2.2 we comment some difficulties observed related to detection of change in slum-like areas in a more detail.

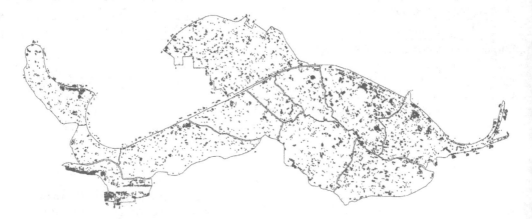

Fig. 11. Identified changes between years 2006 and 2009 (Erdas Imagine). Change pattern enclose changes due to new buildings construction, buildings disappearance and larger rooftop renovations.

### 5.1.3 Kibera population estimation in 2009

Total surface of Kibera was calculated as a sum of housing areas of all 12 villages throughout whole Kibera from vector results obtained with object-based classification. All the vectorized structures have been computed using ArcGIS software and we assumed that all the structures are used for habitat purposes.

Sum of the residential areas show that total residential area of Kibera is around 1.646,883 m², which is 358,706 m² more than total surface from year 1993 given by Sartori et al. (2002), where residential area was 1.333,834 m². We can see that informal settlement population density has increased from year 1993. Since Kibera is surrounded by well-defined residential boundaries and its drastical expansion therefore was not possible, we can assume that the increase of residential area can attribute to denser housing inside the informal settlements through years.

As we can see from Table 5, population estimation of Kibera can vary from 150 thousand up to 650 thousand people living in the informal settlements according to different population density sources. Nevertheless, both limits are high taking into account Kibera is spread on 2.5 km² only.

| Source | Density [people/m²] | People living in Kibera |
|---|---|---|
| MapKibera Project (2008) | 0.0951 | 156,652 |
| IRIN (2006) | 0.2000 | 329,377 |
| AHI US (2005) | 0.3000 | 494,065 |
| Sartori et al. 1 (2002) | 0.3300 | 543,471 |
| Sartori et al. 2 (2002) | 0.3900 | 642,284 |

Table 5. The estimation of population according to the density acquired from different sources. Housing area of Kibera was calculated from results obtained with object classification of GeoEye image from 2009.

## 5.2 Raila village

### 5.2.1 Detailed land cover map of Raila village 2009

We made a detailed classification focused on residential housing only on image subset covering the area of Raila village. The main idea was to automatically obtain polygon shapes of each individual residential object or settlement. Although on VHR image individual objects in general can be distinguished, due to the complexity and variety of dense roof surface there was impossible to extract the shape of every individual residential object or each group of objects (settlements). Roofs and their elements themselves present heterogeneity, resulting in distinct spectral variation within areas of homogeneous land cover. Class „buildings_bright" were possible to detect since they had a very distinct morphological pattern that was contrasted by surrounding (built-up) land cover. Segments of class „buildings_brown" were sometimes integrated with their surroundings (soil) so they could not be readily distinguished. Some urban elements (like rooftops) are a combination of many different surface materials; this produced a spectral response that was difficult to interpret with routine procedures.

Automatic shape detection of each individual residential object would enable good total population estimation. Two approaches to get classified vector shapes closer to individual buildings were considered:

- Manual correction of the shapes of obtained polygons: Small test showed that this is extremely time consuming. Correction took more time than it would take if one would manually digitize the whole image.
- Use of additional attributes that are automatically created in the phase of ENVI EX segmentation: Investigation showed that there is no useful relation within additional attributes of all sub-classes of buildings. Therefore additional attributes could not assist to semi-automatic reclassification of some segments in order to obtain better shapes of individual object in build-up zones.

Finally, focused to residential build-up objects in Raila area four major object classes were modelled and mapped (Fig. 12).

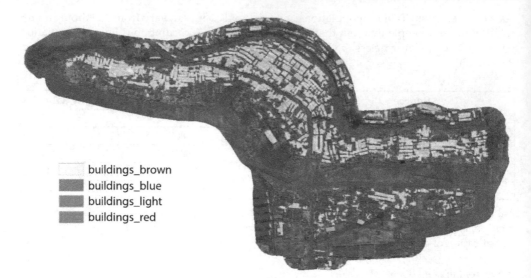

buildings_brown
buildings_blue
buildings_light
buildings_red

Fig. 12. Detailed object based classification results for residential areas in Raila village.

## 5.2.2 Raila village urban growth

Multi-level contextual change detection approach was applied to Raila village to highlight the extent of major urbanisation processes that can take action even in a short time span. As expected the development on southern border of the village is well recognised (Fig. 13) and different time sequences analysed outlined where and when the growth has taken place.

Method incorporates spatial neighbourhood dependence to control the false change information (i.e. inherent changes due to locally based variability in data). As a result it gives a change pattern of positive and negative changes in imagery spectral space. Positive changes correspond to an increase in digital number values at the same location and negative changes to a decrease in spectral values. Method was developed for middle resolution imagery, but its application to Kibera informal settlement data proved this concept to be efficient on VHR imagery as well. In general, majority of objects transformations regardless of their size (small patches to entire buildings) was identified. The results were evaluated with visual control or before-after imagery comparison. Although quantitative assessment was not possible as more detailed independent data were not available, we estimate that more than 90% of changes associated to buildings appearance and disappearance are captured.

Hence, detailed inspection of identified changes outlined several difficulties related to detect change in slum-like areas. First of them is preservation of shape. Because of the use of multi-level information the objects edges may be shrinked to some degree, causing that change is not recognised at entire object extent but only at part of it. Second, the level of change recognition (detection rate) is different in positive and negative spectral space. New materials (i.e. metal plates) used to cover houses have very strong reflection, so difference from low reflecting brown soil to new buildings with bright cover is obvious and unproblematic. In contrast, for example from rusty rooftops to bare soil this difference is not

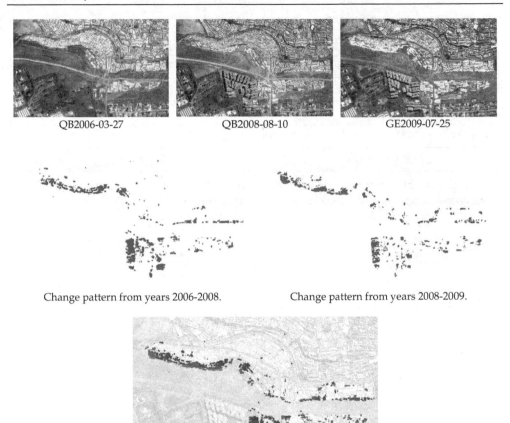

QB2006-03-27                    QB2008-08-10                    GE2009-07-25

Change pattern from years 2006-2008.          Change pattern from years 2008-2009.

Urban growth pattern through years 2006, 2008 and 2009 for Raila village.
Dark red corresponds to 2006-2008 and light red to 2008-2009 period.

Fig. 13. Results of change detecion for urban growth pattern through years 2006, 2008 and 2009 for Raila village, Kibera.

great and consequently buildings disappearance is more difficult to detect. Same applies if new building has dark colour rooftop covers (for example, often used blue colour plates). Main change missing occurrences therefore apply to the just described situations, making difficulties for automatic approaches. To some extent this problem was solved so that the threshold for negative changes was reduced during transformation categorisation step.

### 5.2.3 Raila village population estimation in 2009

According to the total residential area in Raila village obtained from object based classification (section 5.1.1, Table 3), population estimation was calculated based on different density per area parameters (Table 6). Additionaly we illustrate this information also in terms of typical size of houses in the Raila village (Fig. 14).

| Source | Density [people/m²] | Density [people/32m²] | Estimated population of Raila |
|--------|---------------------|------------------------|------------------------------|
| MapKibera Project (2008) | 0.0951 | 3.04 | 4,343 |
| IRIN (2006) | 0.2000 | 6.40 | 9,131 |
| AHI US (2005) | 0.3000 | 9.60 | 13,697 |
| Sartori et al. 1 (2002) | 0.3300 | 10.56 | 15,067 |
| Sartori et al. 2 (2002) | 0.3900 | 12.48 | 17,806 |

Table 6. Density of people living in Raila village according to different sources and total residential area obtained from object-based classification of GeoEye 2009-07-25 image.

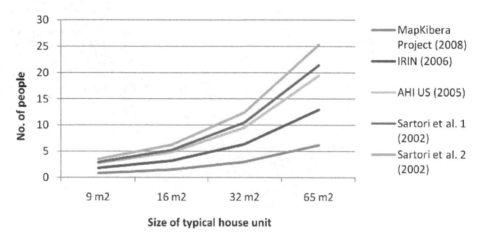

Fig. 14. Density of people according to different sources per typical size of houses in Raila village.

In Table 6 we presented also the density on 32 m² unit since this size of house was measured to be a most typical and frequent (medium range size) housing unit in Raila village. It was also observed that size of small buildings is approximately 16 m², while large buildings reach on average 65 m². Fig. 14 summarise in addition the above relationship.

When perfoming such population estimation it is assumed that all of the build-up area is used for dwelling, which is not allways true as houses may be also used for other activities. For the possible calibration of the above population densities field survey sampling would be required. Nevertheless, Table 6 clearly outlines how population estimation based on critera of simple density per area based modelling can propose up to 4-times higher (different) population estimation.

## 6. Discussion

Informal settlements are a very dynamic phenomenon in space and time and the number of people living in these areas is growing worldwide. The reasons for this are many-sided and were not under detailed examination of this study. Informal settlements represent a particular housing and living conditions which is from a humanitarian point of view in most cases below acceptable standards (UN-HABITAT, 2004; MKP, 2011; Sartory et al. 2002). Due to informal character and low governmental management services in the past, reliable and accurate data about informal settlements and their population is rarely available. On the other side there is a strong need to transform informal into formal settlements and to gain more control about the actual urbanisation progress. Thus, obtaining spatially and temporally accurate information is a first step to establish proper actions in terms of local or regional planning. For these tasks, conventional data sources, such as maps, statistics or even GIS data are usually obsolete, not available, not as accurate as needed or do not hold the information needed (MKP, 2011; Sartory et al. 2002). The case study presented demonstrates how informal settlements can be approached from VHR satellite image data. Using an object based approach of image analysis detailed land cover/use within informal settlements can be obtained to facilitate GIS-based management tasks and population modelling. The application of automatic, even if simple pixel-based, change detection proved to support real-time observation of informal settlement areas whenever appropriate VHR satellite data are available with relatively low processing costs.

Merits of object-based image analysis in dense informal settlements analysis with VHR remote sensing data have been confirmed in several studies (section 2). However some drawbacks still resist. In case of residential land cover/use map derivation main unsolved difficulty is automatic detection and separation of individual houses. Although small differences in heights of rooftops create visually well distinguished boundaries of objects, the heterogeneity of rooftop material and its small scale changeability often overrule the value of neighbourhood houses boundaries relationship. Current limit of object based analysis is also that still requires substantial post-classification routines and check-over that can be done mainly manually through visual control. What characterise this procedure as time demanding whenever geometrically and semantically correct information is aimed.

Change detection applied revealed great potential for long-term monitoring and informal settlements urbanisation growth analysis. Hence more research is needed to provide sufficient detection rate of spectrally lower magnitude changes that are typical for informal settlements specifics and its reflectance intensity representation on satellite imagery. Due to rooftop material used bright (metal) materials are unproblematic to distinguish from soils, however dark rooftop materials (blue, brown colour) are spectrally closer to bare, unpaved (brown) or vegetated (green) soils. Here object based approaches would prove better option as sub-object attributes could be explored and used.

Valuable population estimation can be made with a relatively low cost if residential area if accurately estimated from high-resolution images, although some considerations exist. Area based population estimation model can be used for the informal settlements in other images of similar resolution knowing the number of people living per surface unit. Zhang (2002) exposed some problems of selection of the scales of remote sensing imagery, reduction of influence of plant cover on remote sensing data, stability of the correlation between population and remote sensing indicator variables and correction of building count. In this research we met all mentioned problems in order to accurately estimate population out of VHR satellite data. Nevertheless, if a clear understanding of mentioned issues is considered, reliable population model of population estimation by remote sensing data can be created. Thus, the application of satellite data information (such as accurate information on land use extent and other measures of surface or environmental characteristics) along with socio-economic data may well facilitate complex modelling to estimate population trends.

In terms of remote sensing technology contribution it is necessary to continue to develop new techniques for complex densely packed urban environments such as informal settlements. Emphasis on spectral properties should be considered but also emphasis on the characteristics of the shape, texture, context, and relationship with neighbouring pixels (and/or objects) information needs to be enhanced; as well as integration of the knowledge on corresponding socio-economic drivers should not be neglected.

## 7. Conclusion

Effective methods of monitoring informal settlements are required to generate appropriate data fast enough to assist to local policies and their controlling actions. Remote sensing data are especially powerful in that respect since, apart they are up-to date, they assist to link the geographic location with the accurate socio-economic data.

The results of change detection confirmed that VHR imagery is very promising for immediate monitoring of dense informal residences in the areas where much information is lacking. The results of object-based (contextual) classification of the land use in informal settlements of Kibera were highly accurate, especially if taking into consideration that informal settlements are difficult to be interpreted with automatic or semi-automatic routines. On the other side, the results indicate the problem of the ratio between spectral and spatial heterogeneity of objects in slum-like areas when viewed only from the above (satellite) perspective. Overall, the use of the object-based image analysis holds great promise for dense urban environments and was proved useful for studies of urban change structure and corresponding population estimation.

Satellite derived information can greatly complement the information that is traditionally collected by field observations (UNHCR, 2000). Quantitative information that can be derived from it should not be underestimated. The production of maps with geometrical shapes of settlements can contribute to recover the management of informal settlements, especially when interfaced with database that has information collected on the field. Although several challenges have not been yet solved adequately, e.g. delimitation of individual objects in slum-like areas, we can notice that applications are being developed. Thus (automatic) analysis of objects enables tremendous opportunities for population estimation in informal settlements.

## 8. Acknowledgment

The Slovenian Research Agency has sponsored the post-doctorate research project (Z2-2127). The Centre of Excellence for Space Sciences and Technologies Space-SI is an operation partly financed by the European Union, European Regional Development Fund, and Republic of Slovenia, Ministry of Higher Education, Science and Technology. American Association for the Advancement of Science supported the operation of Map Kibera Trust with donated satellite images of the area.

## 9. References

A History of Kibera (2011). Available from www.brianekdale.com/?p=230.

Addink, E.A., Van Coillie, F.M.B. (Eds.) (2010). *GEOBIA 2010: Geographic Object-Based Image Analysis*, Ghent, Belgium, 29. June - 2. July 2010. ISPRS Vol.No. XXXVIII-4/C7, Archives ISSN No 1682-1777. Available from http://geobia.ugent.be/proceedings/html/papers.html.

AHI US (2005). Affordable housing institute. Available from http://affordablehousinginstitute.org/blogs/us/2005/07/kibera_africas.html.

Barrett, E., Curtis, L. (1986). *Introduction to Environmental Remote Sensing*, pp. 238-241. London: Chapman and Hall.

Benz, U.C., Hofmann, P., Willhauch, G., Lingenfelder, I., Heynen, M. (2004). Multi-resolution, object-oriented fuzzy analysis of remote sensing data for GIS-ready information. *ISPRS Journal of Photogrammetry and Remote Sensing*, Vol. 58, pp. 239-258.

Blaschke, T., Burnett, C., Pekkarinen, A. (2004). New contextual approaches using image segmentation for object-based classification. In: *Remote Sensing Image Analysis: Including the spatial domain*, De Meer, F. and de Jong, S. (Eds.), Kluver Academic Publishers, Dordrecht, pp. 211-236.

Blaschke, T., Lang, S., Hay G.J. (eds.) (2008). *Object-Based Image Analysis: Spatial Concepts for Knowledge-Driven Remote Sensing Applications* (Lecture Notes in Geoinformation and Cartography). Springer.

Dare, P.M., Fraser, C.S. (2001). Mapping informal settlements using high resolution satellite imagery. In: *Int. J. of Remote Sensomg*, Vol. 22, No. 8 pp. 1399-1301.

Davis, M. (2006). The Planet of Slums.

Deer, P. (1998). *Digital change detection in remotely sensed imagery using fuzzy set theory*. Ph.D. thesis, Departments of Geography and Department of Computer Scoence, University of Adelaide, Australia.

Dengsheng, L., Qihao, W., Guiying, L. (2006). Residential population estimation using a remote sensing derived impervious surface approach. *International Journal of Remote Sensing*, Vol. 27, No. 16, pp. 3553-3570.

Galeon F. (2008). Estimation of population in informal settlement communities using high resolution satellite image. *The International Archives of the Photogrammetry: remote sensing and spatial information sciences*, Vol. 37, Beijing.

Giada, S., De Groeve, T., Ehrlich, D. (2003). Information extraction from very high resolution satellite imagery over Lukole refugee camp, Tanzania. *International Journal of Remote Sensing*, Vol. 24, pp. 4251-4266.

Hay, G. J., Blaschke, T., Marceau, D.J., Bouchard, A. (2003). A comparison of three image-object methods for the multiscale analysis of landscape structure. *ISPRS Journal of Photogrammetry & Remote Sensing 57*, pp. 327-345.

Hoffman, P., Strobl, J., Blaschke, T., Kux, H., (2006). Detecting informal settlements from Quickbird data in Rio de Janeiro using an object based approach. Available from http://ispace.researchstudio.at/downloads/2006/165_Hofmann_et_al.pdf

Hofmann, P., 2001. Detecting urban features from IKONOS data using an object oriented approach. *Fist Annual Conference of the Remote Sensing & Photogrammetry Society*, pp. 28-33.

Horton, F. (1974). Remote sensing techniques and urban acquisition. In: *Remote Sensing Techniques for Environmental Analysis,* Estes, J., Senger L. (Eds.) , pp. 243-276. Santa Barbara: Hamilton Publishing Co.

IHC (2007). International Housing Coalition. Urban Investments and Rates of Return: Assessing MCC's Approach to Project Evaluation. Available from http://www.intlhc.org/docs/urban-investment.pdf.

IRIN (2006). Humanitarian news and analysis. A project of the UN Office for the Coordination of Humanitarian Affairs (09/13/2006), "KENYA: Kibera, The Forgotten City". Available from http://www.irinnews.org/Report.aspx?ReportId=62409.

Jianya, G., Haigang, S., Guorui, M., & Qiming, Z. (2008). A review of multi-temporal remote sensing data change detection algorithms. *The International Archives of the Photogrammetry, Remote Sensing and Spatial Information Sciences.*

Johnson, B., Xie, Z. (2011). Unsupervised image segmentation evaluation and refinement using a multi-scale approach. *ISPRS Journal of Photogrammetry & Remote Sensing 66,* pp. 473-483.

Karanja, M. (2010). Myth shattered: Kibera numbers fail to add up. Daily Nation. Available from http://www.nation.co.ke/News/Kibera%20numbers%20fail%20to%20add%20up/-/1056/1003404/-/13ga38xz/-/index.html

Kibera Wikipedia (2011). Available from http://en.wikipedia.org/wiki/Kibera.

Kuffer, M. (2003). Monitoring the Dynamics of Informal Settlements in Dar Es Salaam by Remote Sensing: Exploring the Use of SPOT, ERS and Small Format Aerial Photography. In: *Schrenk, M. (Ed.): Proceedings CORP 2003*, pp. 473-483.

Lang, S., Blaschke, T., Schöpfer, E. (2006). *1st International Conference on Object-based Image Analysis.*
http://www.isprs.org/proceedings/XXXVI/4-C42/papers.htm.

Langford, M., Unwin, D.J. (1994). Generating and mapping population density surfaces within a geographical information system. *The Cartographic Journal*, Vol. 31, No.1, pp.21-26.

Lemma, T. et al. (2005). A Participatory Approach to Monitoring Slum Conditions. Available from http://www.itc.nl/library/papers_2005/conf/sliuzas_par.pdf

Lin, Z., Jin, Y., Li, C. (2001). Urban population geographical information systems. *Proceedings of the 20th International Cartographic Conference*, pp. 1279-1282, Beijing, China, August 6-10, 2001.

Lu, D., Mausel, P., Brondizio, E., Moran E. (2004). Change detection techniques. *International Journal of Remote Sensing 20 (25)*, 2365–2407.

Mason, O.S.., Fraser, C. (1998). Image sources for informal settlement management. In: *Photogrammetric Record*, Vol. 92 (1998), No. 16, pp. 313-330.

MKP (2011). Map Kibera Project. Available from http://mapkibera.org, http://www.mapkiberaproject.org, http://mapkiberaproject.yolasite.com

Netzband, M., Rahman, A. (2010). Remote Sensing for the mapping of urban poverty and slum areas. Panel contribution to the Population-Environment Research Network Cyberseminar, "What are the remote sensing data needs of the population-environment research community?", May 2010.

Radnaabazar, G. et al. (2004). Monitoring the development of informal settlements in Ulanbaatar, Mongolia. http://www.schrenk.at/CORP_CD_2004/archiv/papers/ CORP2004_radnaabazar_kuffer_hofstee.pdf (accessed 2006).

Sartori, G., Nembrini G. & Stauffer, F. (2002). Monitoring of Urban Growth of Informal Settlements and Population Estimation from Aerial Photography and Satellite Imaging. http://www.thirstycitiesinwar.com/files/informalsettlements.pdf

Singh, A. (1989). Digital change detection techniques using remotely-sensed data. *International Journal of Remote Sensing 10*, pp. 989–1003.

Sliuzas, R.V. and Kuffer, M. (2008). Analysing the spatial heterogeneity of poverty using remote sensing : typology of poverty areas using selected RS based indicators. In: *Remote sensing : new challenges of high resolution, EARSeL*, joint workshop, 5-7 March 2008 Bochum, Germany / ed. by C. Jürgens. Bochum : EARSeL, 2008. pp. 158-167.

Sutton, P. (1997). Modeling population density with night-time satellite imagery and GIS. *Computers, Environment and Urban Systems*, Vol. 21, No.3-4, pp. 227-244.

UN-Habitat (2004). Africa on the Move. An urban crisis in the making. Available from http://www.unhabitat.org/downloads/docs/4626_83992_GC%2021%20Africa%2 0on%20the%20Move.pdf

UNHCR (2000). Population estimation and registration. In: *Handbook for Emergencies*, 2nd edition, pp. 118–131, Geneva: UNHCR Publications.

UNHCR (2010). IDP Population Assessment of the Afgooye Corridor – January 2010.

UNHSP (2005). United Nations Human Settlements Programme (UN-HABITAT), Nairobi. Situation analysis of informal settlements in KISUMU, Kenya slum upgrading programme. Available from http://www.unhabitat.org, http://www.unhabitat.org/pmss/listItemDetails.aspx?publicationID=2084

Veljanovski, T. (2008). The problem of false (non-intrinsic) changes in pixel-based change detection on Landsat imagery. *Geodetski vestnik, Vol. 52, No. 3*, pp. 457-474.

Veljanovski, T., Oštir, K. (2011). Influence of radiometric pre-processing on image properties and comparability consistency. *International Conference on Sensors and Models in Photogrammetry and Remote Sensing: SMPR2011*, May 18-19, Tehran-Iran.

Wang, F.Z. (1990). Urban population estimation – multispectral imagery analysis. *Urban Environment and Urban Ecology*, pp. 34-42.

Wang H.H. (1990). City population estimation method with satellite imagery. *Remote Sensing technology*, pp. 48-54.

Zhang, B. (2003). Application of remote sensing technology to population estimation. *Chinese geographical science*, Vol. 13, No.3, pp.267-271.

# 11

# Demonstration of Hyperspectral Image Exploitation for Military Applications

Jean-Pierre Ardouin[1], Josée Lévesque[1],
Vincent Roy[1], Yves Van Chestein[1] and Anthony Faust[2]
*[1]Defence R&D Canada, Valcartier*
*[2]Defence R&D Canada, Suffield*
*Canada*

## 1. Introduction

Airborne hyperspectral imagers have been available from various providers for many years and their performance keeps improving. On the other hand, space-based hyperspectral sensors have only been available from few exploratory missions such as NASA Hyperion on EO-1 (Pearlman et al, 2003) and ESA CHRIS on Proba (Cutter et al, 2003). In recent years, there have been many civilian space missions being planned in different countries (Buckingham & Staenz, 2008), as well as military space demonstrations (Cooley et al, 2006).

Given the increase in potential space-based hyperspectral sensors, Defence R&D Canada (DRDC), which is part of the Canadian department of National defence, began in 2005 a project to demonstrate the military utility of space-based reflective hyperspectral imagery (0.4-2.5 microns) to the Canadian Forces (CF). The project is called HYperspectral iMage EXploitation (HYMEX) and ended its activities in 2010 (Ardouin et al, 2007).

Before the HYMEX project, DRDC had been conducting and sponsoring R&D in the area of hyperspectral image exploitation for a number of years to explore its various possibilities (Davenport & Ressl, 1999; Sentlinger et al, 2003; Webster et al, 2006). The focus of this work was on military target detection applications. In parallel with these activities, the Canadian remote sensing community has also been active in developing hyperspectral applications for various civilian applications related to forestry, agriculture, fisheries, mineral exploration and environmental monitoring (Buckingham et al, 2002). Many hyperspectral techniques developed for civilian applications can be applied to military applications such as terrain characterization.

Building on previous efforts at DRDC and with support from Canadian industry, academic institutions and other government departments, the HYMEX project identified a set of applications and related algorithms to be demonstrated to the Canadian Forces.

This chapter presents an overview of the project, beginning with a description of its main activities (Section 2.0), including field trials, data analysis and algorithms evaluation and the development of an image exploitation software. Then, for each application areas, target detection (Section 3.0), land mapping (Section 4.0) and marine mapping (Section 5.0), we discuss some of the most promising algorithms and show examples of application of these algorithms.

## 2. HYMEX applications and activities

The HYMEX project addressed applications divided in 3 categories:

1.  Target detection and identification: This includes targets such as military vehicles, camouflages and various man-made materials.
2.  Land mapping applications: This includes the characterization of soil and vegetation.
3.  Marine mapping applications: This includes beach characterization and near-shore bathymetry, as well as water color mapping.

As mentioned in the introduction, DRDC developed expertise and advanced exploitation techniques for target detection over the years but had limited expertise for terrain analysis and water mapping. To identify available techniques for the last 2 categories, HYMEX conducted a survey of about 60 different groups from industry, academia and other government departments forming the Canadian remote sensing community. Most algorithms from the groups that responded have been included in the project.

The project delivered a suite of near-operational hyperspectral image exploitation tools called HOST (Hyperspectral Operational Support Tools). HOST is developed in the IDL language as an add-on to a commercial-off-the-shelf package called ENVI available from ITT Visual Information Solutions. ENVI is already a widely used platform for hyperspectral image analysis. HOST builds onto ENVI to provide tools specifically oriented for the needs of the Canadian Forces. The HYMEX System Integrator responsible for the development of HOST was MacDonald Dettwiler and Associates (Richmond, BC).

HYMEX adopted an algorithms validation strategy that led to the selection of the best algorithms to be integrated into HOST and explore the performance limits of these algorithms. Knowing these limits is an important element of information for transitioning the tools into operational use by the Canadian Forces. The validation strategy included the following components:

*Field Trials:* The HYMEX project conducted several field trials at different locations across Canada to acquire remotely sensed hyperspectral imagery and ground truth for testing the various algorithms. Each location had specific characteristics likely to affect the performance of the algorithms. Marine trials were conducted in two different locations (East and West coast). Land trials were conducted in an Acadian forest (CFB Gagetown, NB), a boreal forest (CFB Valcartier, QC), an aspen dominant forest (CFB Wainwright, AB) and in prairie grassland (CFB Suffield, AB). One trial was conducted in winter under snow conditions. While some space-based hyperspectral imagery was acquired, the project main data type came from airborne hyperspectral sensors. Airborne data contains similar information as space-based sensors, and can be used in the interim to evaluate algorithms. In fact, even though HYMEX focused on demonstrating the utility of space-based imagery, the results can easily be transposed into the utility of airborne sensors.

*Data analysis:* All HYMEX algorithm providers analyzed trial data using performance metrics approved by DRDC. In addition to providing quantitative performance evaluation and comparisons of the algorithms, the overall data analysis allowed the demonstration in a limited way of a) how the performance varies in different environments, b) the advantage of hyperspectral (HSI) over multispectral (MSI) imagery and c) the effect of varying ground sampling distance (GSD) via the analysis of airborne data at different altitudes or of different sensors.

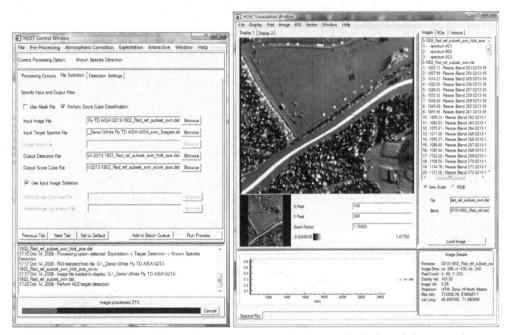

Fig. 1. HOST two main windows: the visualization window on the left and the control window on the right

*HOST demonstrations*: We developed the HOST software in three successive iterations. Each iteration ended with a live or hands-on demonstration of the tools to CF stakeholders and image analysts. Feedback from the demonstration participants were integrated in the following iterations and the selection of the algorithms to be integrated for the iteration was based on interim data analysis results. As explained above, HOST is an add-on to ENVI. While ENVI is a powerful exploitation package for advanced hyperspectral imagery users, HOST is oriented towards military end-users with introductory knowledge of hyperspectral image exploitation. In order to present a simplified and more uniform user interface than ENVI, HOST is organized in two main windows as illustrated in Figure 1: the visualization window and the control window. The HOST visualization window regroups, in a single window, many of the familiar visualization tools offered by ENVI (image display, plot display, available bands list, region of interest tool and the vector layer manipulation tool). By regrouping these tools in a single window, the user can more easily keep track of these functions as they are applied to specific hyperspectral images. The control window provides, in a single window, an interface to the parameters of the different advanced exploitation algorithms offered by HOST. The user would typically use the control window to setup batch scripts to process many hyperspectral images without user intervention. The HOST control window is organized in different logical categories of algorithms such as: pre-processing, atmospheric correction, exploitation and interactive tools. A Navigation Tool also allows loading customized task descriptions with links to the HOST user interface. This guides the user through the algorithms needed to accomplish a task and the selection of the parameters for those algorithms.

## 3. Target detection algorithms

Throughout the HYMEX project, DRDC gained experience in applying algorithms for target detection applications. In this section, we describe a typical processing chain (atmospheric correction, detection and target abundance estimation (Roy, 2010)) used in the project and present results from an experiment aimed at evaluating the performance of the target abundance estimation part of the processing chain using data collected in difficult illumination and atmospheric conditions.

In late October and early November 2009, DRDC collected airborne hyperspectral imagery near Suffield, Alberta (50°13′N, 110°10′W) using an Itres SASI-600 SWIR pushbroom imaging system. The sensor was flown at various altitudes ranging from 330m to 1700m above ground level in order to acquire imagery at across-track ground sampling distance (GSD) of 0.4m, 1.0m, and 2.0m, while along-track GSD remained constant at 1.0m. Imagery was collected between 13h00 and 15h00 local time, which resulted in sun elevation between 17 and 25 degrees. Furthermore, thin altostratus clouds and an overcast of altocumulus clouds on the first (29 Oct) and second day (03 Nov) of collect respectively degraded the illumination conditions considerably, as illustrated in Figure 2. Compared to typical reflective hyperspectral field trials usually conducted under clear skies and high solar elevation, this data collection was conducted under significantly adverse environmental conditions not often considered in the hyperspectral literature.

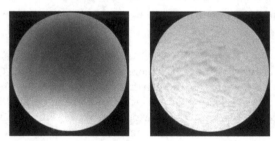

Fig. 2. Typical sky conditions on 29 Oct 2009 (left) and 03 Nov 2009 (right).

One objective of this field trial was the evaluation of the constrained energy minimisation (CEM) algorithm (Settle, 2002) sub-pixel abundance estimation accuracy. For this purpose, we designed targets of known abundances made of thin strips of painted metal, as illustrated in Figure 3 below. The design allowed changes to the abundance level by varying the distance between the strips of metals while their overall size (5m x 5m) ensured that they filled completely at least one pixel in the imagery, as showed in Figure 4. We used two different types of paint to vary the contrast between the target and background, one beige (see Figure 3) and one green (not shown). The base color was mixed with small quantities (2 to 10% per volume) of black feature-less paint to control the spectral features depth and overall signature albedo. A total of 6 targets were used in this experiment.

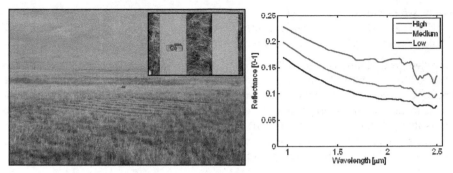

Fig. 3. Left) Example of controlled abundance target. Right) Spectral signatures of the beige paint at different albedo levels, as measured in field conditions using an ASD FieldSpec Pro spectrometer.

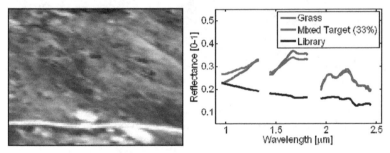

Fig. 4. Left) SWIR 3-colors composite of targets, imaged at a 2.4m GSD (coarsest resolution). Right) Background (red) and target (blue) signatures as measured by the airborne sensor, compared to the target library signature (black).

DRDC favours automated and adaptive approaches to hyperspectral target detection, minimizing user interaction and processing time as much as possible. In this context, we used the following processing chain for this dataset exploitation:

1. We first manually identified and removed 12 out of the original 100 spectral bands due to their low signal to noise level. We also removed unusable part of the imagery due to sensor vignetting. This is typically done only once for a given sensor.

2. We then converted the imagery from at-aperture radiance to apparent reflectance units using the empirical line method (Smith & Milton, 1999), with five very large targets (greater than 10x10 pixels) of known reflectance. This was possible due to the controlled environment of this imagery collection; else we would typically have used a semi-empirical technique such as the QUick Atmospheric Correction (QUAC) method (Bernstein et al, 2005).

3. The CEM algorithm requires a description of the background 2nd order statistics (mean and covariance). For this calculation, we identified a subset of background pixels from the complete image using the RX anomaly detection algorithm (Reed & Yu, 1990) by keeping only the first 90% lowest scoring pixels. Second order statistics were then evaluated using this limited set of pixels.

4.  Finally, we calculated the score for all pixels of the image using the CEM algorithm. As described in Settle (2002), when properly normalised, the CEM output is an estimation of the searched target signature abundance in the pixel under test.

The controlled ground targets were collected in 45 different images. On 29 Oct, the abundance was set to 50% while on 03 Nov it was reduced to 33%. All images were manually interpreted to delineate the area of the targets in the images. The CEM scores were averaged over each target area to derive an "average abundance", as shown in Figure 5. This was necessary because as imaged, the targets had inhomogeneous abundance over their physical extent, particularly at the finest GSDs. This suggests that the target design could be improved for future experiments by using thinner strips of material more closely spaced together.

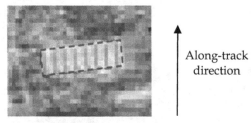

Along-track direction

Fig. 5. Example of manual delineation of target area. Target is 5m x 5m, GSD is 0.4m (across-track) by 1.0m (along track).

Estimated abundance error was calculated using the root mean square error (RMSE) and the estimation bias, both normalized by the true abundance in order to get a relative error in percent:

$$\text{Relative RMSE [\%]} = \frac{100\%}{\alpha_{\text{true}}} * \sqrt{\frac{\sum \left( \alpha_{\text{evaluated}} - \alpha_{\text{true}} \right)^2}{N}} \tag{1}$$

$$\text{Relative bias} = \frac{100\%}{\alpha_{\text{true}}} * \frac{1}{N} \sum \left( \alpha_{\text{evaluated}} - \alpha_{\text{true}} \right) \tag{2}$$

RMSE results are presented in Table 1. Overall, the root mean square error for this experiment is between 11.7% and 30%. In absolute terms, the overall RMSE translates to 0.064 and 0.078 for the 0.5 and 0.33 abundance targets respectively. The retrieved abundances were slightly underestimated, with bias of -1.9% and -14.7% again on the 0.5 and 0.33 abundance targets respectively. Since atmospheric conditions degraded between the two collects, it is unclear if the observed increase in error is related to the change in illumination conditions, to the lower abundance level considered, or to a combination of both.

The results achieved are encouraging and show that target abundance can be retrieved at the subpixel level using the CEM algorithm with a high accuracy. The fact that the estimated abundances are generally lower than the true abundances which is consistent with an error that could have been introduced during the manual delineation of targets area, by assigning larger areas to targets than their true area. Also, the imaging system true point spread function has not been characterized and taken into account in this analysis; non-uniform sampling over the GSD could lead to an underestimation of the sub-pixel abundances (Settle, 2004).

| Target type | | Target configuration | |
|---|---|---|---|
| Contrast type | Albedo type | $\alpha = 0.5$ (29 Oct 2009) | $\alpha = 0.33$ (03 Nov 2009) |
| High (beige) | High | 11.7 | 30.0 |
| High (beige) | Medium | 13.4 | 24.3 |
| High (beige) | Low | 12.8 | 19.6 |
| Low (green) | High | 11.9 | 20.8 |
| Low (green) | Medium | 13.2 | 24.5 |
| Low (green) | Low | 13.2 | 21.7 |
| Average over all targets: | | 12.7 | 23.5 |

Table 1. Relative RMS errors of the retrieved abundances using the CEM algorithm; $\alpha$ denotes the target abundance.

The results demonstrate the robustness of the processing chain; with minimal user interaction and using a simple processing chain suitable for near real-time exploitation, targets can be characterized at the sub-pixel level even under adverse illumination conditions. This demonstrates the processing chain's military utility, and indicates that it could be adapted to the detection and characterization of spectral signatures of interest in a military operational context.

## 4. Land mapping applications

Land mapping applications were studied in collaboration with the University of New-Brunswick, the University of Alberta, the University of Lethbridge, York University and Laval University. The work was oriented towards soil and vegetation characterization and mapping for trafficability and environmental applications. In Section 4.1, algorithms for classification and the extraction of vegetation canopy attributes (density, structure) were evaluated using airborne hyperspectral data acquired over three Canadian Forces bases (CFB). The resulting validated hyperspectral products were then used to improve a trafficability model developed by the University of New Brunswick for Gagetown military base as well as promote environmentally sustainable training on military bases. Winter airborne images were also acquired over the Montmorency experimental forest (near Quebec City) to investigate the potential of winter imagery to better derive forest information. In Section 4.2 we show that among the classification algorithms that were evaluated, the Mercury algorithm (an evidential-reasoning-based supervised classification algorithm developed by the University of Lethbridge (Peddle & Ferguson, 2002)) achieved the best performance. Finally, Section 4.3 shows results from a laboratory study conducted by the University of Alberta demonstrating how hyperspectral techniques can be used to discriminate between vegetation stresses caused by exposure to different toxic industrial chemicals (Rogge et al, 2008).

### 4.1 Trafficability and the monitoring of military training areas

This section presents results obtained for the two main land applications of HYMEX, trafficability and the monitoring of training ranges to promote environmentally sustainable training. Table 2 provides a summary of the trials and the objectives sought for each application. The primary objective of each trial was the validation of algorithms used to derive vegetation cover information such as type, density and height, the presence of wetlands and the determination of soil type. These surface features are easily derived from hyperspectral imagery and can contribute to improve knowledge of the terrain for the purposes of trafficability and environmental applications. Each trial was conducted in a different vegetation background ranging from various forest biomes (deciduous, mixte, boreal) to prairie grassland. Details regarding each trial, the available ground truth and the algorithms used to analyse the various datasets can be found in (Ardouin et al, 2007).

| Trial location | Date | Sensor | GSD | Background | Objectives | Application |
|---|---|---|---|---|---|---|
| CFB Gagetown, NB | Sep 2005 | Probe-1 | 15m | Acadian forest | Forest parameter algorithms validation (mixed deciduous) & CFB Gagetown trafficability model | trafficability & sustainable training |
| CFB Wainwright, AB | Sep 2006 | AISA | 4m | Boreal/grassland | Forest parameter algorithms validation (single deciduous species) | trafficability |
| Montmorency Experimental Forest, QC | Jun 2004 Feb 2007 | AISA | 4m 4m | Boreal/summer Boreal/winter | Forest parameter algorithms validation (mixed conifers), wetland mapping & summer/winter dataset investigation | trafficability |
| CFB Suffield, AB | Sep 2006 | AISA | 4m | Prairie grassland | Map invasive species, burnt areas, soil disturbance | sustainable training |

Table 2. HYMEX land mapping application trials

**CFB Gagetown trial**. One of the objectives of this trial was to improve the trafficability model used by the Army Meteorological Center (AMC) at CFB Gagetown to plan training exercises, avoid erosion by vehicles and promote environmentally sustainable training. The model currently use as input, the soil moisture content simulated by the University of New Brunswick (UNB) Forest Hydrology Model (ForHyM2) which is based on air and ground temperatures, soil type, the amount of precipitation and the wind speed and direction. Improvement of the trafficability model was achieved by the addition of above ground restrictions such as the forest type (hard/softwood), density and height which can be readily derived from hyperspectral remote sensing. Figure 6b shows a vegetation species classification derived from 15m GSD imagery collected by the Probe-1 sensor (Figure 6a). The overall accuracy (81.8%) and Kappa coefficient (0.78) are based on 533 pixels. These results were obtained with the University of Lethbridge Mercury classification algorithm. Figure 7 shows the shadow fraction of the forest canopy which was derived from spectral mixture analysis (SMA) (Peddle & Smith, 2005) along with the sunlit deciduous fraction, the sunlit conifer fraction and the background fraction. The image shadow fraction was found to correlate the best with LAI, as measured on the ground on 29 plots with hemispherical pictures ($R^2$ = 0.55) with an average difference between the SMA LAI and the ground LAI of less than 0.5 LAI. Figures 8a and 8b show two forest canopy structures, stem density and stand height, as output from the University of Lethbridge Multiple Forward Mode 3-D Canopy Reflectance Model (MFM-3D) applied to the modified geometric optical mutual shadowing model (GOMS) (Peddle et al, 2003). MFM-3D uses a Look-up-tables (LUT) approach based on various ranges and increments of forest structure parameters (density, horizontal & vertical crown radius, crown height and height distribution) as input. The ranges and increments can be determined either from field data or automatically without prior knowledge. Inversion of MFM-3D model produces results when image reflectance values match the modelled reflectance. Field and MFM-3D stand height produced less than 2 m average height difference with the under-estimation of the MFM-3D model attributed to the difficulty in locating neighbouring pixels with similarity to the center pixel.

The addition of above ground restrictions to UNB trafficability model, as determined by the vegetation layers described above, helps produce more refined trafficability classes as illustrated in Figure 9. Figure 10a shows the graphical user interface (GUI) of the route planning tool. Once all the available layers are loaded into the Input Dialog, the user can select from the Interactive Parameters which restrictions to apply for a particular vehicle type. Examples of route planning for four types of military vehicles are shown in Figure 10b with a low environmental concern (not avoiding areas with a high rutting index), and 10c with a high environmental concern (avoiding areas prone to produce ruts).

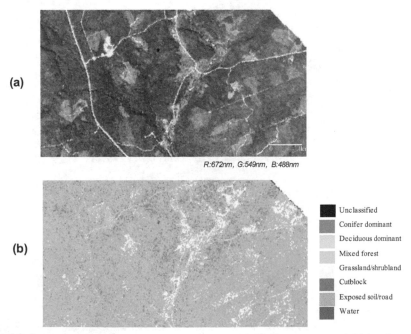

(a)

R:672nm, G:549nm, B:488nm

(b)

- Unclassified
- Conifer dominant
- Deciduous dominant
- Mixed forest
- Grassland/shrubland
- Cutblock
- Exposed soil/road
- Water

Fig. 6. (a) Sub-image of Probe-1 airborne imagery of CFB Gagetown and (b) landcover classification derived from an evidential reasoning classifier with classes of vegetation and exposed soil and roads. Overall accuracy: 81.8%. Kappa coefficient: 0.78 (D. Peddle, U. of Lethbridge).

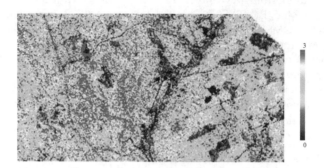

Fig. 7. Sub-image of Probe-1 airborne imagery of CFB Gagetown showing the shadow fraction of the forest canopy derived from spectral mixture analysis (D. Peddle, U. of Lethbridge).

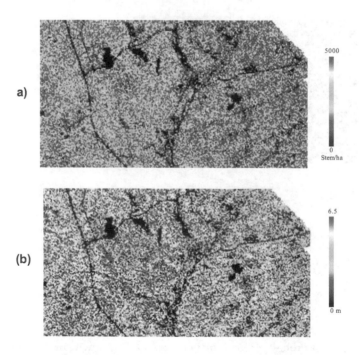

Fig. 8. Sub-image of Probe-1 airborne imagery of CFB Gagetown showing forest structural parameters (a) stand density and (b) stand height as derived from the MFM-3D model (D. Peddle, U. of Lethbridge).

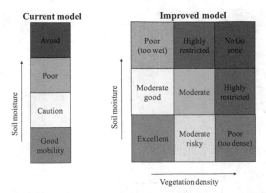

Fig. 9. Off-road trafficability classes for the current (left) and improved models (right).

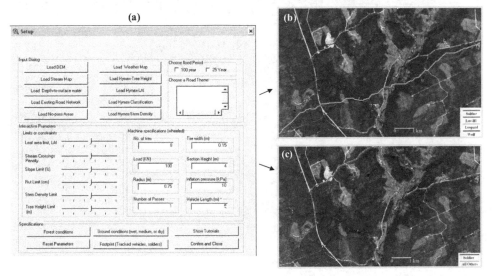

Fig. 10. (a) Route planning for four military vehicles with (b) low and (c) high environmental concern.

**CFB Wainwright trial.** A similar trafficability model was also constructed by UNB for CFB Wainwright with the objective to demonstrate its portability to different soil and vegetation ecosystems. The above ground trafficability is determined by a semi-arid prairie landscape with well defined dry and wet seasons. The vegetation is composed of grassland with areas of deciduous trees (aspen, balsam poplar and willow) and the topography is gentle. The hyperspectral vegetation products made available to UNB are the land cover classes and the leaf area index (LAI) (Figures 11b and 11c) provided by York University under contract to HYMEX. Figures 11d to 11f show the optimal route planning between point A and point B using a Wolf and a LAV vehicles when constrained by wet areas and vegetation during the wet season (Figure 11d), when constrained by wet areas and vegetation during the dry season (Figure 11e) and when constrained by wet areas and vegetation during the wet season and the requirement to move along tree lines as closely as possible (Figure 11f).

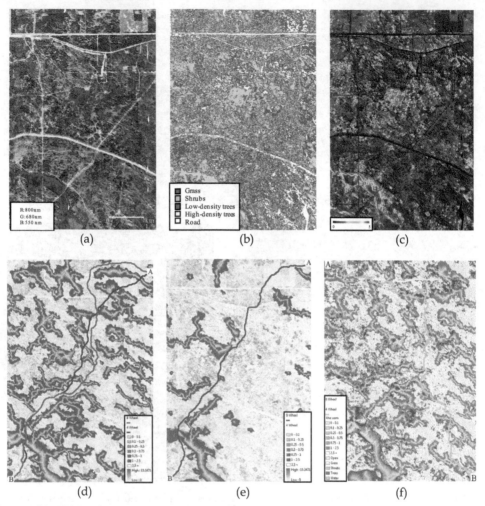

Fig. 11. (a) AISA color composite, (b) land cover classes and (c) leaf area index map of CFB Wainwright (York University). Route between A and B for a Wolf and a LAV vehicles: (d) when constrained by wet season conditions, (e) when constrained by dry season conditions and (f) when constrained by wet season conditions and the requirement to move along tree lines (UNB).

**Montmorency Experiment Forest (MEF) trials.** Two airborne hyperspectral datasets are available for the MEF site, one from Jun 2004 (summer) and one from Feb 2007 (winter). The objective for imaging this site was to test algorithms for vegetation mapping in a coniferous dominant forest ecosystem. The winter 2007 dataset was acquired for the purpose of investigating the usefulness of summer/winter data to extract relevant terrain information for trafficability in the boreal forest. One first attempt to address this later objective is to map tamarack trees (*Larix laricina*) which can be used as an indicator species for the location of peatlands areas dominated by trees, a wetland type being of interest for trafficability.

Tamarack trees thrive in open areas because of their intolerance to shade and their resentment to compete with other species (Beeftink, 1951). They also adapt very well to poorly drained soil. Their presence is generally associated with peatlands although their absence do not indicate that there is no wetlands. Spectrally, tamaracks are similar to other conifers in summer and to deciduous trees in winter because they loose their leaves (needles) before winter. A combination of summer and winter data allows the exploitation of this unique characteristic of the tamarack trees to locate and map treed wetlands. Figure 12b shows a RGB of AISA data acquired in winter 2007 at the MEF. Figure 12c is a moisture stress vegetation index in which the red color represents exposed bark. When looking closer at the single tree where the red arrow is pointing in Figure 12a, it is easy to recognise from the shape of its shadow on the snow that it is a defoliated coniferous tree. Not having this information on hand it would be difficult to determine whether the trees in the red color class in Figure 12c are deciduous, dead spruce or fir, or dormant tamarack.

Under contract to DRDC, Laval University applied four filters, each made of a band ratio index and a predefined threshold, to classify the tamarack trees in the AISA image with 95% of the tamarack pixels correctly classified and only 1.2% of the remaining pixels misclassified as hardwood trees. Each filter discriminates tamarack trees from other forest features such as other softwood and deciduous. The effects of the application of the first 3 and the first 4 filters are shown in Figure 13a and 13b.

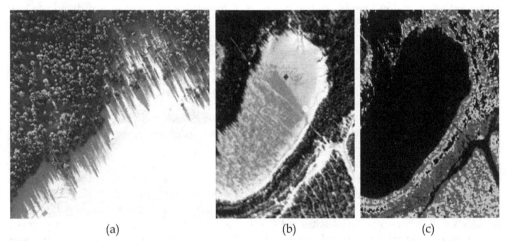

(a)                                    (b)                                    (c)

Fig. 12. (a) high resolution color image, (b) AISA color composite (R:1290 nm , G:1655 nm, B:2189 nm), and (c) a moisture stress vegetation index from the AISA winter image of an area at the Montmorency experimental forest. The red arrow points at a tamarack tree.

**CFB Suffield trial.** The objectives of this trial were primarily environmentally oriented. Despite a semi-arid climate, prairie grasslands are very sensitive to the introduction of invasive species which are often dispersed during military training and along the maintenance roads of pipelines and gas wells. Leafy spurge and crested wheat are the main invasive species and can easily spread in windward direction into preserved native prairie areas. Moreover, there is a need for monitoring training areas for an environmentally sustainable training. This is to ensure that excessive training does not over stress the soil and

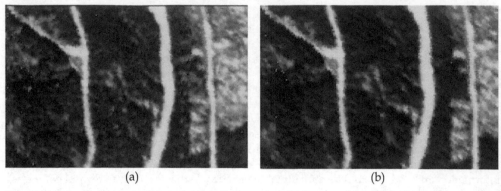

(a)                                                    (b)

Fig. 13. (a) result from applying the first 3 filters in which several hardwood pixels are misclassified as tamarack, and (b) result from applying the first 4 filters (red dots are correctly classified tamarack pixels). (Prof. Sylvie Daniel and Gaël Briant, Laval University).

therefore the vegetation capacity to recover. The imaged areas include a wide range of soil and vegetation species including invasive species, burned areas, cultivation and grazing areas, wetlands, and various levels of disturbances by vehicle pathways. At the time of the airborne hyperspectral survey (Sep 2006), the prairie landscape was dry and with the exception of the low lands and around wetlands, the vegetation exhibited a low photosynthetic activity which resulted in less pigment absorption in the visible and more apparent absorption features in the short wave infrared by other plant cell constituents such as lignin and cellulose. The following results obtained by the University of Alberta (under contract for HYMEX) demonstrate the potential of this dataset for mapping soil and vegetation at CFB Suffield to help the environmentally sound planning of military exercises.

Soil was determined to have a high clay content. Thus, exposed soil was mapped using the spatial distribution of the depth of the clay absorption feature in the vicinity of 2200 nm after removing the vegetation effect using an orthogonal subspace projection and known green and dry vegetation endmember spectra (Figure 14b). Band depth was measured using the continuum removal between 2210nm and 2230 nm. The band depth was classified into four classes (Figure 14c) defined as (1) low clay absorption depth (green) corresponding to natural undisturbed terrain, (2) slightly (yellow) and strongly (blue) disturbed soils areas and (4) high clay absorption (red) which correspond to bare soils, active roads, non-vegetated dry wetlands and burnt areas. In undisturbed grassland areas the soil is covered with dry grass, old grass residue and a layer of moss. When the surface is disturbed, some of the soil becomes exposed and the amount of moss and old residues decreases. Thus, a good indicator of vegetation recovery following exercises would be a dominance of dry grass. Figure 14d shows an RGB of the clay band depth (red), the most dominant grass endmember (green) and an endmember associated with spectra of moss covered soil measured with a field spectrometer (blue). These three classes of endmembers can easily be associated with the following three conditions: (1) permanently disturbed areas such as roads and areas surrounding gas wells (red), (2) recently disturbed areas where the moss and old residues are removed (green) and (3) undisturbed areas covered by moss, old residues and grass (blue).

Invasive species could not be spectrally identified due to the overall dryness of the vegetation cover. The RGB composite of Figure 15b displays the most abundant green

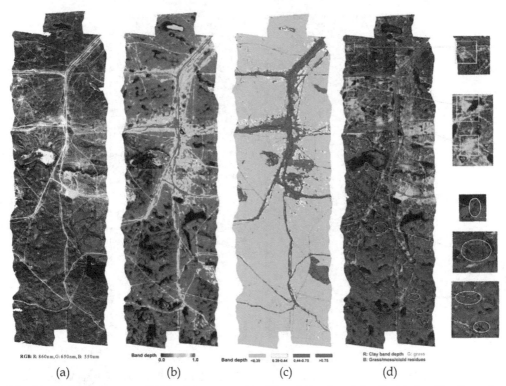

RGB: R: 860nm, G: 650nm, B: 550nm     Band depth 0.0 — 1.0     Band depth <0.39  0.39-0.44  0.44-0.75  >0.75     R: Clay band depth  G: grass
B: Grass/moss/oil/oil residues

(a)                    (b)                    (c)                    (d)

Fig. 14. (a) AISA imagery RGB color composite, (b) soil clay band depth, (c) classification of soil clay band depth and (d) RGB of the clay band depth (red), the most dominant grass endmember (green) and moss covered soil spectra (blue). (B. Rivard, U. of Alberta)

vegetation endmember in red and the two dominant dry vegetation endmembers in green and blue. The green vegetation (Red color) is located in low land areas which are often located in the vicinity of wetlands. Local cattle grazing is allowed in some area of the military base. The dry vegetation shown in blue represents overgrazed areas which can be compared to impacted areas from training exercise in other area of the military base, thus showing the potential for environmental monitoring for sustainable training. The black area in the northern part represents a recently burnt area where vegetation hasn't started to grow back.

## 4.2 Classification algorithms comparison

As indicated in the previous section, the Mercury supervised classification algorithm (Peddle & Ferguson, 2002) performed well during HYMEX field trials and as a result was integrated into the HOST software. In order to evaluate its performance in more details, we compared Mercury to all the supervised classification algorithms offered by the ITT's ENVI 4.8 software (Van Chestein, 2011). Two data sets were used, a 15 m spatial resolution Probe-1 hyperspectral image of CFB Gagetown and a 4 m resolution AISA hyperspectral imagery of CFB Wainwright. The classes defined for each dataset are listed in Table 3.

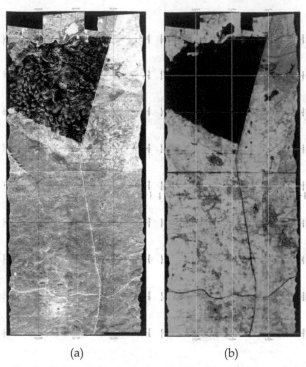

(a)                                          (b)

Fig. 15. (a) RGB "true color" (red: 640nm, green: 550nm, blue: 460nm) AISA image. (b) RGB composite image of the most widely spread green vegetation endmember (Red) and two dominant dry vegetation endmembers (green, blue). (B. Rivard, U. of Alberta).

| Class | Train | Test |
|---|---|---|
| dirt road | 15 | 15 |
| paved road | 52 | 50 |
| conifer forest | 140 | 62 |
| deciduous forest | 263 | 113 |
| grass | 170 | 80 |
| clearcut area | 143 | 58 |
| water | 106 | 39 |

(a)

| Class | Train | Test |
|---|---|---|
| water | 49 | 49 |
| trees | 91 | 98 |
| shrubs | 68 | 65 |
| muskeg | 36 | 50 |
| gravel | 40 | 35 |
| grass | 33 | 37 |

(b)

Table 3. Number of training and test pixels for each class: (a) classes for CFB Gagetown data (Probe-1, 15m GSD) and (b) classes for CFB Wainwright data, (AISA, 4m GSD).

During the comparison, it was found that the tested algorithms behaved differently as the number of bands used in the classification process increases. Some see classification accuracy increase, others prove unaffected by the number of bands while a third group see the accuracy decrease, albeit slightly.

An immediate advantage in using fewer bands is that processing times are shorter, which is very convenient when analyzing large files. It can also be useful to identify which algorithms are most consistent in accuracy as the number of bands is changed. This way, by using a classification algorithm with known consistency, the optimal band-set can be selected quickly after performing a few tests.

The major finding was that the Mercury algorithm consistently provides very high overall classification accuracy values as illustrated in Table 4. It proves stable and offers the advantage of not requiring that the number of training pixels for each class be at least equal to the number of bands used plus one as is the case with the Maximum Likelihood and Mahalanobis Distance techniques. Mercury's accuracy increased with the number of bands and it offered the highest individual accuracy values in both datasets. Using Mercury on the principal components yielded lower accuracy than with the original dataset. With the Maximum Likelihood algorithm applied to the principal components, results were almost identical to those obtained with the original data. The following table illustrates the findings.

| Number of bands → | 4 | 7 | 20 | 50 | Average |
|---|---|---|---|---|---|
| Mercury | 82.30% | 83.60% | 88.20% | 91.30% | 86.4% |
| Support Vector Machine | 84.40% | 83.10% | 84.40% | 85.60% | 84.4% |
| Mahalanobis Distance | 77.60% | 75.20% | 86.20% | 89.50% | 82.1% |
| Neural Network | 81.60% | 84.20% | 88.30% | 65.20% | 79.8% |
| Maximum Likelihood | 87.60% | 73.00% | 77.40% | 75.20% | 78.3% |
| Minimum Distance | 80.30% | 76.70% | 76.60% | 76.50% | 77.5% |
| Parallelepiped | 75.60% | 74.00% | 77.40% | 77.00% | 76.0% |
| Spectral Angle | 67.20% | 67.60% | 68.80% | 69.00% | 68.2% |
| Spectral Information Divergence | 68.70% | 66.60% | 68.40% | 67.60% | 67.8% |
| Binary Encoding | 52.00% | 51.40% | 56.60% | 66.70% | 56.7% |

Table 4. Comparison of Mercury and ENVI supervised classification algorithms accuracy. Green is for the algorithm that ranked 1st in classification accuracy and yellow is the algorithm that ranked second.

The study also highlighted the fact that class accuracy varies greatly with the choice of bands in most algorithms. Figures ranging from 0% to 100% accuracy were observed in some algorithms but Mercury came out with very consistent global figures.

In summary, Mercury compares very favourably with ITT's offering for global and class accuracy and for all algorithms, one would be well advised to run a few tests as to the number and choice of bands to ensure optimal feature accuracy.

## 4.3 Chemical effects on vegetation

### 4.3.1 Plants as chemical detectors

The Canadian Centre for Mine Action Technologies initiated a study in 2003 to investigate the possibility of exploiting advances in genetic engineering and plant biotechnology to design a process by which plants, local to a region of interest, could be genetically modified

(GM) to be sensitive to the compounds known to permeate the soil around emplaced landmines. In this case it was envisioned that the plant's genes would also be designed to include a reporting mechanism, signalling the presence of these compounds through a change in the plant's structure, appearance or some other physical characteristic. The Deyholos group at the University of Alberta was funded to conduct the initial study (Deyholos et al, 2006).

At the same time, the United States' Defense Advanced Research Projects Agency (DARPA) inititated the Biological Input Output Systems (BIOS) program. The BIOS program's objective was to produce basic biochemical modules for future use in plant or microbial-based detectors of chemical and biological compounds of strategic interest. Collaboration between the two projects advanced efforts in developing a human-readable biological signalling event (Deyholos et al, 2007; Antunes et al, 2006)

The DARPA-funded team at Colorado State University went on to develop the first generation plant-based sensor capable of detecting 2,4,6-TNT in the low ppt (parts per trillion) range. The Canadian effort made significant progress in the development of a root-to-shoot transducer system and an effective visual reporter system (Deyholos, 2009).

This effort clearly demonstrated that plants' natural responses to chemicals in their environment could be harnessed, exploited and enhanced to provide an *in situ* chemical detection capability of remarkable sensitivity. This observation, amongst others, led to a study to investigate whether it might be possible to detect, through optical means, the naturally occurring effects of exposure to various chemical agents on vegetation, by which *in situ* vegetation may provide a highly sensitive stand-off detection capability to chemical exposures occurring at ground level. These agents cause stress and damage to surrounding vegetation the extent of which is dependent on dosage and time of exposure.

### 4.3.2 Passive detection

It is well recognized that reflective hyperspectral imagery (400-2500nm) is well suited to analyze vegetation. Under the Canadian Space Agency HERO program, a feasibility study was conducted (Peddle et al, 2008) to determine whether a space-based system such as HERO can be used to detect toxic industrial chemicals indirectly by detecting the stress that these chemical cause on vegetation. Recognizing that this could have a potential military application, we pursued this project under HYMEX by conducting a laboratory evaluation of the stresses caused by various chemicals.

The aim of this investigation was to provide information that would help quantify the potential of reflective hyperspectral imagery for chemical and biological surveillance, reconnaissance involving plants exposed to Toxic Industrial Chemicals and Materials (known as TICs and TIMs, such as Ammonia, Sulphur Dioxide, Chlorine, Hydrogen Sulphide, Hydrogen Cyanide, Cyanides, Phosgene).

The two objectives of this study were to determine if: 1) vegetation subjected to TICs could be distinguished from background vegetation during varying growth stages (new growth to senescence) and environmental stresses; and, 2) different TICs could be distinguished based on the vegetation spectral response. This work was conducted by teams at the University of Alberta (Rivard et al, 2008).

This research team examined the spectral response of individual leaves of three common Canadian plant species (poplar (Populus deltoides, Populus trichocarpa), wheat (Triticum aestivum), canola (Brassica napus)), which were subjected to fumigation with gaseous phase toxic industrial chemicals and chemicals precursor to chemical warfare agents (e.g. ammonia and sulphur dioxide) (TICs). Treatments were designed to allow quantification of the variation in spectra that might be expected due to environmental, developmental, and stochastic effects on the physiological state of individual plants within each species.

The test plants were grown in controlled environment chambers at the University of Alberta, using standardized conditions. Each spectral measurement collected with the ASD® FR spectrometer, as shown in Figure 16, consisted of an average of 10 scans. Multiple scans were taken per leaf location to reduce the effects of noise. For each leaf, three different locations were measured located approximately halfway between the main leaf vein and the leaf edge, precluding overlap of areas measured. The measurements from each leaf were then averaged accounting for spectral variability across the leaf. For smaller leaves (e.g. new growth) only one or two measurements were possible.

Fig. 16. Basic set-up for spectral measurements. Inset is an image of the ASD® Leaf Clip, the field spectrometer used to collect plant data.

The study was broken into two phases: 1) to capture the spectral variability of the various leaf growth stages (new to senescing leaves) observed in each of the three plant types; and 2) subjecting the plants to environmental stresses (e.g. drought) and the following five industrially relevant gaseous phase TICs: ammonia ($NH_3$), sulphur dioxide ($SO_2$), hydrogen sulphide ($H_2S$), chlorine ($Cl_2$), and hydrogen cyanide (HCN). The experimental data were analyzed to determine if the various treatments resulted in specific leaf spectral features related to TICs. Figure 17 illustrates typical effects of the chemicals on plants and Figure 18 depicts representative spectra collected, in this case for canola exposed to $Cl_2$ and $SO_2$. Here one can see key absorption features observed in endmember spectra, which were exploited in subsequent analysis.

Observations showed that both environmental stress and TIC treatments induce similar spectral features inherent to plants, which can be related primarily to chlorophyll and water loss. These include pigments in the visible and cellulose, lignin, lipids, starches, and sugars in the short wave infrared. Although no specific spectral features could be tied to individual TICs, an analysis of the data using vegetation indices, which focus on key spectral bands associated with chlorophyll, pigments and water content, showed that the TICs and

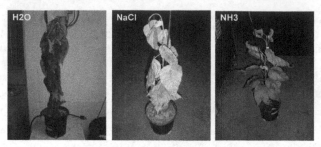

Fig. 17. Examples of environmental ($H_2O$, NaCl) and chemicals ($NH_3$) stresses on plants.

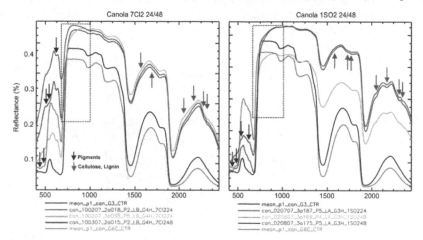

Fig. 18. Representative spectra for canola experiments using $Cl_2$ and $SO_2$. Mean spectra of control G1 (black line, mature healthy) and G4C (pink line, high senescence) are included for reference. Arrows denote key absorption features observed in endmember spectra compared with G1. Black dotted box denotes smoothing of red edge. Chemical exposure differences is apparent.

environmental stresses result in diagnostic light reflectance data trends from healthy mature to highly stressed leaves.

Comparison of relevant vegetation indices, such as that depicted in Figure 19, showed that specific combinations could be used to distinguish $NH_3$, $SO_2$, $Cl_2$ consistently across all three species (Rogge et al, 2008). The trends result from the variable leaf response within plants, between plants and between species and it is expected much of the variability observed within species would be preserved or even enhanced in nature. As such it is encouraging for the possible detection of TIC effects on natural vegetation using airborne/spaceborne imagery.

As the detection methodology was developed from leaf-level observations, it is important to note that field trials remain to be conducted in order to test if the findings of this study can be extended to the detection of TICs in the natural environment. The principal unknown is the effect of varying vegetation canopy structural parameters (e.g. canopy gaps, leaf area) and background properties (litter and soil reflectance) on the specific data trends that were identified.

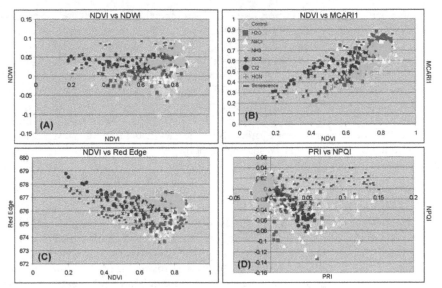

Fig. 19. A selection of vegetation indices across all species for treatments with NaCl, NH₃, SO₂, Cl₂, HCN, and, dehydration (H₂O), senescence and controls plants. The existence of species-specific responses of vegetation to TICs presents both a challenge and an opportunity for regional remote sensing.

While the exact physiological response to each stressor remains to be understood, the existence of species-specific responses of vegetation to TICs presents both a challenge and an opportunity for regional remote sensing.

## 5. Marine algorithms

The HYMEX project studied several potential marine applications in collaboration with Borstad Associates and the Dept. of Fisheries and Ocean (Institut Maurice Lamontagne and Bedford Institute of Oceanography). We conducted airborne hyperspectral surveys on East and West coasts of Canada to evaluate algorithms for near-shore bathymetry, beach trafficability, near-shore bottom type mapping as well as retrieval of chlorophyll and suspended matter concentrations as indicator of water clarity. More details are provided in (Ardouin, 2007). Through this work it was realized that most of these algorithms can be applied to multispectral imagery and that their experimental validation is difficult. The later is particularly true for products that vary with time (or current) and thus would require many measurement stations (for validation) that would need to operate coincidentally with the airborne survey and be distributed over the area of the survey.

More recently, we tasked OEA Technologies to provide an operational assessment of HYMEX marine algorithms. In this assessment, a distinction was made between dynamic (e.g. water color) and static (e.g. bathymetry) products. It was pointed out that the Canadian Forces needs for off-shore dynamic products (e.g. water colour) is already fulfilled by marine multispectral sensors (MERIS, MODIS) with pixel size > 250m (Williams, 2009). There might however be a niche for hyperspectral sensors (airborne and spaceborne) which typically have better spatial resolution (e.g. from submeter to tens of meter) for near-shore static and dynamic products. The better spatial resolution and increased number of bands of hyperspectral sensors might provide an ability to handle the more complex near-shore environment. Potential static products to consider include target detection and near-shore bottom characterization in support of mine countermeasures and battlespace mapping and possibly submarine operations. To this we can also add near-shore bathymetry in support of route survey, battlespace mapping, anti-submarine warfare and submarine operations. While not requiring hyperspectral sensing, HSI could keep playing a role (e.g. selection of optimal bands) in the development of new dynamic products for both near-shore and off-shore applications. Overall, this assessment point to possible follow-up for marine applications development with hyperspectral sensors.

## 6. Conclusion

In this chapter, we discussed a wide variety of military applications resulting from the exploitation of reflective hyperspectral imagery. These applications were demonstrated in the DRDC HYMEX project, allowing DRDC and the Canadian Forces stakeholders to get more familiar with the military utility of hyperspectral imagery. While some of these applications such as target detection are relatively mature and are near to operational deployment, others still require further development but are representative of the unique capability of hyperspectral remote sensing. The many datasets that were acquired and the algorithms and exploitation tools that were developed in the project are being used to continue the development of hyperspectral technology at DRDC. One avenue that is being pursued is the development of an airborne hyperspectral real-time target detection demonstration system. We are also looking at opportunities to further develop the land mapping and marine applications areas as well as potential space-based demonstration with international partners.

## 7. References

Antunes, M.S.; Ha, S.B.; Tewari-Singh, N.; Morey, K.J.; Trofka, A.M.; Kugrens, P.; Deyholos, M. & Medford J.I. (2006) A synthetic de-greening gene circuit provides a reporting system that is remotely detectable and has a re-set capacity, *Plant Biotechnology J.*, Vol. 4, No. 6, (November 2006), pp. 605-622

Ardouin, J.-P.; Lévesque, J. & Rea, T.A. (2007) A Demonstration of Hyperspectral Image Exploitation for Military Applications, *Proc. of the 10th International Conference on Information Fusion (FUSION 2007)*, Quebec, Canada, (9-12 July 2007), pp. 1-8., ISBN 978-0-662-45804-3

Beeftink, H.H. (1951) Some observations on tamarack or eastern larch. *Forestry Chronicle*, Vol. 27, No. 1, (1951), pp. 38-39

Bernstein, L.S.; Adler-Goldstein, S.M.; Sundberg, R.L., Levine, R.Y.; Perkins, T.C.; Berk, A.; Ratkowski, A.J.; Felde, G. & Hoke, M.L. (2005) A new method for atmospheric correction and aerosol optical property retrieval for VIS-SWIR multi- and hyperspectral imaging sensors : QUAC (QUick Atmospheric Correction), *IEEE International Geoscience and Remote Sensing Symposium*, (25-27 July 2005), pp. 3549-3552, ISBN 0-7803-9050-4

Buckingham, R.; Staenz, K. & Hollinger (2002) A., Review of Canadian Airborne and Space Activities in Hyperspcetral Remote Sensing, *Canadian Aeronautics and Space Journal*, Vol. 48, No. 1, (2002), pp. 115-121

Buckingham, R. & Staenz, K. (2008) Review of current and planned civilian space hyperspectral sensors for EO, *Can. J. Remote Sensing*, Vol. 34, Suppl. 1, (2008), pp. S187-S197

Cooley T.; Davis, M. & Straight, S. (2006) ARTEMIS-Advanced Tactically-Effective Military Imaging Spectrometer: Tactical Satellite 3 for Responsive Space Missions, *International Symposium on Spectral Sensing Research*, Bar Harbor, MN, (May 2006)

Cutter, M.A.; Johns, L.S.; Lobb, D.R.; Williams, T.L. & Settle, J.J. (2003) Flight Experience of the Compact High Resolution Imaging Spectrometer (CHRIS), *Proc. of SPIE Imaging Spectrometry IX*. Vol. 5159, (2003) pp. 392-405

Davenport, M. & Ressl, W. (1999) Shadow Removal in Hyperspectral Imagery, *International Symposium on Spectral Sensing Research (ISSSR 99)*, Las Vegas (November 1999)

Deyholos, M.; Faust, A.A.; Miao, M.; Montoya, R. & Donahue, D.A. (2006) Feasibility of landmine detection using transgenic plants, in *Detection and Remediation Technologies for Mines and Minelike Targets XI*, J. Thomas Broach, Russell S. Harmon, John H. Holloway, Jr, eds., *Proc. SPIE*, Vol. 6217, (May 2006), pp. 700-711

Deyholos, M.K.; Rogge, D.; Rivard, B. & Faust, AA. (2007) Plants as sensors for toxic industrial chemicals and munitions: A feasibility analysis, *In Vitro Cellular & Developmental Biology-Animal*, Vol. 43, pp. S7-S7.

Deyholos, M. (2009) *Demonstration of Plant-based Explosives Detection*, Contract Report Defence R&D Canada – Suffield, DRDC Suffield CR-2010-010 (2009).

Pearlman, J.S.; Barry, P.S; Segal, C.C.; Shepanski, J.; Beiso & D. & Carman, S.L. (2003) Hyperion, a Space-Based Imaging Spectrometer, *IEEE Trans. on Geoscience and Remote Sensing*, Vol. 41, No. 6, (June 2003), pp. 1160- 1173, ISSN 0196-2892

Peddle, D.R. & Ferguson, D.T. (2002) Optimization of Multisource Data Analysis using Evidential Reasoning for GIS Data Classification, *Computers & Geosciences*, Vol. 28, No 1, (2002), pp. 45-52

Peddle, D.R.; Franklin, S.E.; Johnson, R.L.; Lavigne, M.B. & Wulder, M.A. (2003) Structural Change Detection in a Disturbed Conifer Forest Using a Geometric Optical Reflectance Model in Multiple-Forward Mode. *IEEE Trans. on Geoscience and Remote Sensing* Vol. 41, No. 1, (Jan 2003), pp. 163-166, ISSN 0196-2892

Peddle, D.R. & Smith, A.M. (2005) Spectral Mixture Analysis of Agricultural Crops: Endmember Validation and Biophysical Estimation in Potato Plots, *International Journal of Remote Sensing*, Vol. 26, No. 22, (2005), pp. 4959-4979

Peddle, D.R.; Boulton, R.B.; Pilger, N.; Bergeron, M. & Hollinger, A. (2008) Hyperspectral detection of chemical vegetation stress: evaluation for the Canadian HERO satellite mission, *Can. J. Remote Sensing*, Vol. 34, Suppl. 1, (2008), pp. S198-S216

Reed, I. S. & Yu, X. (1990) Adaptive multiple-band CFAR detection of an optical pattern with unknown spectral distribution, *IEEE Trans. on Acoustics, Speech, and Signal Processing*, Vol. 38, No. 10, (October 1990), pp. 1760-1770, ISSN 0096-3518

Rivard, B.; Deyholos, M. & Rogge, D. (2008) *Chemical Effects on Vegetation Detectable in optical bands 350-2500 nm*, Contract Report Defence R&D Canada - Suffield , DRDC Suffield CR-2008-234, (2008)

Rogge, D.; Rivard, B.; Deyholos, M.; Lévesque, J. & Faust, A. (2008) Toxic Industrial Chemical Effects on Poplar, Canola, and Wheat detectable over the 450-2500 nm Spectral Range, *Intl. Geoscience & Remote Sensing Symposium (IGARSS 2008)*, Boston, MA, (July 2008)

Roy, V. (2010) Hybrid algorithm for hyperspectral target detection, *Proc of SPIE Algorithms and Technologies for Multispectral, Hyperspectral, and Ultraspectral Imagery XVI*, Vol. 7695, May 2010, pp. 1-10

Sentlinger, G.; Davenport M., & Ardouin, J.-P. (2003) Automated Target Recognition in Hyperspectral Imagery using Subpixel Spatial Information, *Proc. SPIE Aerosense – Automatic Target Recognition XIII*, Vol 5094, (April 2003)

Settle, J. (2002) On constrained energy minimization and the partial unmixing of multispectral images, *IEEE Trans. on Geoscience and Remote Sensing*, Vol. 40, No. 3, (March 2002), pp. 718-721, ISSN 0196-2892

Settle, J. (2004) On the Use of Remotely Sensed Data to Estimate Spatially Averaged Geophysical Variables, *IEEE Trans. on Geoscience and Remote Sensing*, Vol. 42, No. 3, (March 2004), pp. 620-631, ISSN 0196-2892

Smith, G. M. & Milton, E.J. (1999) The use of the empirical line method to calibrate remotely sensed data to reflectance, *International Journal of Remote Sensing*, Vol. 20, (1999), pp. 2653-2662

Van Chestein, Y. (2011) *Comparative evaluation of the Mercury classification algorithm: On the influence of the number of bands on classification accuracy using hyperspectral data*, Defence R&D Canada - Valcartier Technical Memorandun, TM 2010-385, (March 2011)

Webster, A.H.; Davenport, M.R. & Ardouin, J.-P. (2006) 3D Deconvolution of Vibration Corrupted Hyperspectral Images, *International Symposium on Spectral Sensing Research 2006*, Bar Harbor, MA., (May 2006).

Williams, D.; Vachon, P.W; Wolfe, J.; Robson, M.; Renaud, W.; Perrie, W.; Osler, J.; Isenor, A.W.; Larouche, P.; Jones, C. (2009) *Spaceborne Ocean Intelligence Network – SOIN – fiscal year 08/09 year-end summary*, Defence R&D Canada - Ottawa External Client Report, DRDC Ottawa ECR 2009-139, (September 2009)

# Permissions

The contributors of this book come from diverse backgrounds, making this book a truly international effort. This book will bring forth new frontiers with its revolutionizing research information and detailed analysis of the nascent developments around the world.

We would like to thank Boris Escalante-Ramírez, for lending his expertise to make the book truly unique. He has played a crucial role in the development of this book. Without his invaluable contribution this book wouldn't have been possible. He has made vital efforts to compile up to date information on the varied aspects of this subject to make this book a valuable addition to the collection of many professionals and students.

This book was conceptualized with the vision of imparting up-to-date information and advanced data in this field. To ensure the same, a matchless editorial board was set up. Every individual on the board went through rigorous rounds of assessment to prove their worth. After which they invested a large part of their time researching and compiling the most relevant data for our readers. Conferences and sessions were held from time to time between the editorial board and the contributing authors to present the data in the most comprehensible form. The editorial team has worked tirelessly to provide valuable and valid information to help people across the globe.

Every chapter published in this book has been scrutinized by our experts. Their significance has been extensively debated. The topics covered herein carry significant findings which will fuel the growth of the discipline. They may even be implemented as practical applications or may be referred to as a beginning point for another development. Chapters in this book were first published by InTech; hereby published with permission under the Creative Commons Attribution License or equivalent.

The editorial board has been involved in producing this book since its inception. They have spent rigorous hours researching and exploring the diverse topics which have resulted in the successful publishing of this book. They have passed on their knowledge of decades through this book. To expedite this challenging task, the publisher supported the team at every step. A small team of assistant editors was also appointed to further simplify the editing procedure and attain best results for the readers.

Our editorial team has been hand-picked from every corner of the world. Their multi-ethnicity adds dynamic inputs to the discussions which result in innovative outcomes. These outcomes are then further discussed with the researchers and contributors who give their valuable feedback and opinion regarding the same. The feedback is then collaborated with the researches and they are edited in a comprehensive manner to aid the understanding of the subject.

Apart from the editorial board, the designing team has also invested a significant amount of their time in understanding the subject and creating the most relevant covers. They scrutinized every image to scout for the most suitable representation of the subject and create an appropriate cover for the book.

The publishing team has been involved in this book since its early stages. They were actively engaged in every process, be it collecting the data, connecting with the contributors or procuring relevant information. The team has been an ardent support to the editorial, designing and production team. Their endless efforts to recruit the best for this project, has resulted in the accomplishment of this book. They are a veteran in the field of academics and their pool of knowledge is as vast as their experience in printing. Their expertise and guidance has proved useful at every step. Their uncompromising quality standards have made this book an exceptional effort. Their encouragement from time to time has been an inspiration for everyone.

The publisher and the editorial board hope that this book will prove to be a valuable piece of knowledge for researchers, students, practitioners and scholars across the globe.

# List of Contributors

**Hyun Jung Cho**
Department of Integrated Environmental Science, Bethune-Cookman University, Daytona Beach, FL,USA

**Deepak Mishra**
Department of Geosciences, Mississippi State University, USA
Northern Gulf Institute and Geosystems Research Institute, Mississippi State University, MS State, MS, USA

**John Wood**
Harte Research Institute for Gulf of Mexico Studies, Texas A&M University-Corpus Christi, Corpus Christi, TX, USA

**Ameris Ixchel Contreras-Silva, Alejandra A. López-Caloca and F. Omar Tapia-Silva**
Centro de Investigación en Geografía y Geomática "Jorge L. Tamayo" A.C., CentroGeo, Mexico

**F. Omar Tapia-Silva**
Universidad Autónoma Metropolitana, Unidad Iztapalapa, Mexico

**Sergio Cerdeira-Estrada**
Comisión Nacional para el Conocimiento y, Uso de la Biodiversidad, CONABIO, Mexico

**Shrinidhi Ambinakudige and Kabindra Joshi**
Mississippi State University, USA

**Milena Andrade**
Federal University of Pará, Amazon Advance Studies (NAEA), Brazil

**Claudio Szlafsztein**
Federal University of Pará, Center of Environment (NUMA), Brazil

**Seyed Kazem Alavipanah, Somayeh Talebi and Farshad Amiraslani**
Faculty of Geography, Department of Cartography, University of Tehran, Tehran, Iran

**Jing Peng and Chaojian Shi**
Shanghai Maritime University, P.R. China

**Dimitrios D. Alexakis, Athos Agapiou and Diofantos G. Hadjimitsis**
Cyprus University of Technology, Department of Civil Engineering and Geomatics, Cyprus

**Apostolos Sarris**
Foundation for Research and Technology, Institute for Mediterranean Studies, Laboratory of Geophysical, Satellite Remote Sensing and Archaeoenvironment, Greece

**Kayembe wa Kayembe Matthieu**
University of Lubumbashi, D.R. Congo

**Mathieu De Maeyer and Eléonore Wolff**
Université Libre de Bruxelles, Belgium

**Satoshi Suzuki and Takemi Matsui**
Kansai University, Tokyo Metropolitan University, Japan

**Tatjana Veljanovski, Urša Kanjir, Peter Pehani and Krištof Oštir**
Scientific Research Centre of the Slovenian Academy of Sciences and Arts, Slovenia

**Tatjana Veljanovski, Peter Pehani and Krištof Oštir**
Space-SI – Centre of Excellence for Space Science and Technologies, Slovenia

**Primož Kovačič**
Map Kibera Trust, Kenya

**Jean-Pierre Ardouin, Josée Lévesque, Vincent Roy and Yves Van Chestein**
Defence R&D Canada, Valcartier, Canada

**Anthony Faust**
Defence R&D Canada, Suffield, Canada